DMV Seminar
Band 15

Birkhäuser Verlag
Basel · Boston · Berlin

Albert H. Schatz
Vidar Thomée
Wolfgang L. Wendland

Mathematical Theory of Finite and Boundary Element Methods

1990

Birkhäuser Verlag
Basel · Boston · Berlin

Authors:

Albert H. Schatz
Dept. of Mathematics
Cornell University
Ithaca, N.Y. 14853
USA

Vidar Thomée
Dept. of Mathematics
Chalmers University of Technology
University of Göteborg
S-41296 Göteborg
Sweden

Wolfgang L. Wendland
University of Stuttgart
Mathematisches Institut A
Pfaffenwaldring 57
D-7000 Stuttgart 80
Federal Republic of Germany

The seminar was made possible through the support of the
Stiftung Volkswagenwerk

Deutsche Bibliothek Cataloguing-in-Publication Data

Mathematical theory of finite and boundary element methods /
A. H. Schatz ; V. Thomée ; W. L. Wendland. – Basel ; Boston ;
Berlin : Birkhäuser, 1990
(DMV-Seminar ; Bd. 15)
ISBN 3-7643-2211-X (Basel ...)
ISBN 0-8176-2211-X (Boston)
NE: Schatz, Albert H. [Mitverf.]; Thomée, Vidar [Mitverf.]; Wendland,
Wolfgang L. [Mitverf.]; Deutsche Mathematiker-Vereinigung:
DMV-Seminar

ISBN 3-7643-2211-X
ISBN 0-8176-2211-X

Contents

Preface

These are the lecture notes of the seminar "Mathematische Theorie der finiten Element- und Randelementmethoden" organized by the "Deutsche Mathematiker-Vereinigung" and held in Düsseldorf from 07. - 14. of June 1987.

Finite element methods and the closely related boundary element methods nowadays belong to the standard routines for the computation of solutions to boundary and initial boundary value problems of partial differential equations with many applications as e.g. in elasticity and thermoelasticity, fluid mechanics, acoustics, electromagnetics, scattering and diffusion. These methods also stimulated the development of corresponding mathematical numerical analysis. I was very happy that A. Schatz and V. Thomée generously joined the adventure of the seminar and not only gave stimulating lectures but also spent so much time for personal discussion with all the participants.

The seminar as well as these notes consist of three parts:

I. An Analysis of the Finite Element Method for Second Order Elliptic Boundary Value Problems by A. H. Schatz.

II. On Finite Elements for Parabolic Problems by V. Thomée.

III. Boundary Element Methods for Elliptic Problems by W. L. Wendland.

The prerequisites for reading this book are basic knowledge in partial differential equations (including pseudo-differential operators) and in numerical analysis. It was not our intention to present a comprehensive account of the research in this field, but rather to give an introduction and overview to the three different topics which shed some light on recent research.

These lecture notes could have been ready two years ago if I were not the bad slow guy. The more grateful I am to A. Schatz, V. Thomée and the Birkhäuser Verlag for their patience: I am happy that the notes now are available. I want to thank the DMV and in particular my colleague G. Fischer for encouraging and organizing the seminar; the "Volkswagen-Stiftung" for their generous support; and the hosts of 'Schloß Mickeln' for their great efforts in making our stay a most pleasant one. I am espescially indebted to A. Schatz and V. Thomée for their wonderful contributions to the seminar and to all participants for their interest, the lively discussions and their good company.

Stuttgart, Mai 1990. Wolfgang Wendland

An Analysis of the Finite Element Method for Second Order Elliptic Boundary Value Problems. An Introduction

Alfred H. Schatz
Cornell University
Ithaca, N.Y. U.S.A.

0. Introduction, Table of Contents.

These notes represent an expansion of lectures given at Dusseldorf in the summer of 1987. It is hoped that they will provide an introduction to the mathematical analysis of the finite element method for second order elliptic boundary value problems. They are geared towards students who have a good background in analysis but are not familiar with the finite element method. Thus we start at the beginning with the aim of understanding the method and developing basic techniques and ideas which seem to be fundamental to it. The treatment here is neither exhaustive nor general. Rather, topics are covered which we feel address important problems whose solutions make use of basic techniques. We have attempted to give self-contained proofs wherever possible. It is hoped that these will be adequate in order that the student be able to understand some recent results in the field. For the convenience of the student a bibliography has been included which covers some topics not treated in these notes.

I would like to thank Bruce Wade and Adam Lutoborski for carefully reading these notes, Lars Wahlbin for interesting conversations and Arletta Havlik for her excellent typing job.

An outline of these notes is as follows:

<u>Section 1</u>: Some function spaces, notation and preliminaries.

<u>Section 2</u>: Some finite element spaces and their properties.

<u>Section 3</u>: Orthogonal projections onto finite element spaces in L_2, H^1 and H_0^1.

<u>Section 4</u>: Galerkin finite element method for second order elliptic boundary value problems. Basic H^1 and L_2 estimates.

<u>Section 5</u>: Indefinite second order elliptic problems.

<u>Section 6</u>: Local error estimates.

<u>Section 7</u>: An introduction to grid refinement. An application to boundary value problems with non-convex corners.

<u>Section 8</u>: Maximum norm estimates for the L_2 projection. A method using weighted norms.

<u>Section 9</u>: Maximum norm estimates for the Galerkin finite element method for second order elliptic problems.

1. Some Function Spaces, Notation and Preliminaries.

For simplicity, throughout these lecture notes, Ω will be a bounded domain in $\mathbb{R}^2$ which is either polygonal or has a smooth boundary. In this section we shall discuss the properties of various function spaces which are defined on Ω. We shall prove some Poincare inequalities which will be used to find equivalent norms on certain Sobolev spaces. These results will be used in Section 2 to prove some basic approximation properties of finite element spaces.

1A. Some Function Spaces.

Let $\alpha = (\alpha_1, \alpha_2)$ be a multi-index and $x = (x_1, x_2) \in \mathbb{R}^2$. For $m \geq 0$ an integer, $C^m(\overline{\Omega})$ will denote the space of real valued functions having m continuous derivatives up to order m on $\overline{\Omega}$. This is a Banach space with the norm

$$\|u\|_{C^m(\Omega)} = \sum_{|\alpha| \leq m} \max_{x \in \overline{\Omega}} |D^\alpha u(x)|.$$

If $m = 0$ we set $C^0(\overline{\Omega}) = C(\overline{\Omega})$. The single bar $|\cdot|_{C^m(\overline{\Omega})}$ will be used to denote the semi-norm

$$|u|_{C^m(\overline{\Omega})} = \sum_{|\alpha| = m} \|D^\alpha u\|_{C(\overline{\Omega})} .$$

Furthermore, the notation $C^m(\overline{\Omega}) = C^m$ will be used when the domain Ω is clear.

For $m \geq 0$ an integer, $H^m(\Omega)$ (respectively $H_0^m(\Omega)$) is the Hilbert space which is the completion of $C^\infty(\overline{\Omega})$ (respectively $C_0^\infty(\overline{\Omega})$) with respect to the norm corresponding to the inner product

$$(u,v)_{m,\Omega} = \sum_{|\alpha|\leq m} (D^\alpha u, D^\alpha v)_\Omega .$$

Here $(\cdot,\cdot)_\Omega$ denotes the usual L_2 inner product

$$(u,v)_\Omega = \int_\Omega uv dx .$$

Hence the $H^m(\Omega)$ norm is given by

$$\|u\|_{m,\Omega} = \left[\sum_{|\alpha|\leq m} \|D^\alpha u\|_\Omega^2 \right]^{1/2} ,$$

where

$$\|u\|_\Omega = \left[\int_\Omega u^2 dx \right]^{1/2} .$$

The single bars $|\cdot|_{m,\Omega}$ will be used to denote the semi-norms

$$|u|_{m,\Omega} = \left[\sum_{|\alpha|=m} \|D^\alpha u\|_\Omega^2 \right]^{1/2} .$$

<u>Remark 1.1</u>. Equivalently, $H^m(\Omega)$ may be characterized as the space of functions in $L_2(\Omega)$ having distributional derivatives up to order m which are in $L_2(\Omega)$. More generally $W^{m,p}(\Omega)$, $1 \leq p \leq \infty$ is the space of functions in $L_p(\Omega)$ having distributional derivatives up to order m which are in $L_p(\Omega)$.

For $u \in H^1(\Omega)$ we set

$$\nabla u = \left(\frac{\partial u}{\partial x_1}, \frac{\partial u}{\partial x_2}\right),$$

the gradient of u, and

$$\|\nabla u\|_\Omega = |u|_{1,\Omega}.$$

1B. Some Poincare Inequalities and Equivalent Norms.

In the following two lemmas we shall prove some special cases of some useful Poincare inequalities. We only remark that Lemma 1.1 can be proved for more general domains than are considered. However it will suffice for our needs.

Lemma 1.1. Let Ω be either a square in $\mathbb{R}^2$ with edges of length $d > 0$ or the triangle with vertices at $(0,0)$, $(d,0)$, $(0,d)$. Then

i) if $u \in H^1(\Omega)$,

$$\|u\|_\Omega^2 \le d^2\|\nabla u\|_\Omega^2 + d^{-2}\left[\int_\Omega u dx\right]^2. \tag{1.1}$$

ii) If $u \in C^1(\bar{\Omega})$,

$$\|u\|_{C(\bar{\Omega})} \le C(d\|\nabla u\|_{C(\bar{\Omega})} + d^{-2}|\int_\Omega u dx|). \tag{1.2}$$

where C is independent of u and d.

Proof: Since $C^1(\bar{\Omega})$ is dense in $H^1(\Omega)$, it suffices to consider $u \in C^1(\bar{\Omega})$ in both (1.1) and (1.2). Let us first prove (1.2). Let $x_0 \in \bar{\Omega}$ be such that $\sup_{x\in\bar{\Omega}}|u(x)| = |u(x_0)|$. If $u(x_0) = 0$, (1.2) is trivial. In the case that $u(x_0) > 0$ and $y \in \bar{\Omega}$, it follows from the mean value theorem (since both domains are convex) that

$$0 \leq u(x_0) - u(y) \leq \sqrt{2}\, d\|\nabla u\|_{C(\bar{\Omega})} .$$

Integrating with respect to y_1 and then y_2 we obtain

$$u(x_0) \leq C(d\|\nabla u\|_{C(\bar{\Omega})} + d^{-2}|\int_\Omega u(y)dy)|).$$

If $u(x_0) < 0$, then by a similar argument we arrive at

$$-u(x_0) \leq C(d\|\nabla u\|_{C(\bar{\Omega})} + d^{-2}|\int_\Omega u(y)dy|).$$

Taken together, these last two inequalities imply (1.2). To prove (1.1) in the case of the square, we use the identity

$$(u(x)-u(y))^2 = \left[\int_{y_1}^{x_1} \frac{\partial u}{\partial \zeta}(\zeta,y_2)d\zeta + \int_{y_2}^{x_2} \frac{\partial u}{\partial \eta}(x_1,\eta)d\eta\right]^2,$$

and the Schwarz inequality to obtain

$$(1.3) \qquad u^2(x) + u^2(y)$$

$$\leq 2d\left[\int_0^d(\frac{\partial u}{\partial \zeta}(\zeta,y_2))^2 d\zeta + \int_0^d (\frac{\partial u}{\partial \eta}(x,\eta))^2 d\eta\right] + 2u(x)u(y).$$

Integrating, it follows that

$$2d^2\|u\|_\Omega^2 \le 2d^4\|\nabla u\|_\Omega^2 + 2\left[\int_\Omega u dx\right]^2,$$

which completes the proof of (1.1) in the case of a square. In the case of a right triangle we reflect u as an even function across the hypotenuse. The extended function is continuous on the square and piecewise C^1. The inequality (1.3) holds for the extended function on the square and (1.1) follows from the fact that u was extended as an even function.

As an immediate consequence of Lemma 1.1 we have the following equivalence of norms on $H^m(\Omega)$:

<u>Lemma 1.2</u>. Let Ω be as in Lemma 1.1 with $d = 1$. If $m \ge 1$ is an integer, there exist constants C_1 and $C_2 > 0$ such that for all $u \in H^m(\Omega)$

$$(1.4) \qquad C_1\|u\|_{m,\Omega}^2 \le |u|_{m,\Omega}^2 + \sum_{|\beta|\le m-1} \left[\int_\Omega D^\beta u dx\right]^2 \le C_2\|u\|_{m,\Omega}^2.$$

<u>Proof</u>: The right hand inequality is obtained by applying Schwarz's inequality to each of the terms $\left[\int_\Omega D^\beta u dx\right]^2$. The left hand inequality can be proved by induction. For $m = 1$ we apply (1.1) directly to $\|u\|_\Omega$ to obtain (1.4). Suppose now that (1.4) holds for $m-1$, then

$$\|u\|_m^2 = |u|_{m,\Omega}^2 + |u|_{m-1,\Omega}^2 + \sum_{|\beta|\le m-1} \left[\int_\Omega D^\beta u dx\right]^2.$$

The inequality (1.4) now easily follows by applying (1.1) to all derivatives $D^\beta u$ of order $|\beta| = m-1$, which completes the proof.

Let $\mathcal{P}_K(\Omega)$ denote the space of polynomials of degree $\le K$ on Ω. A consequence of (1.4) is that the norm on the quotient space $H^m(\Omega)/\mathcal{P}_{m-1}(\Omega)$ is equivalent to the semi-norm $|u|_{m,\Omega}$.

<u>Lemma 1.3</u>. Let Ω be as in Lemma 1.2. Then there exist constants C_1 and $C_2 > 0$ such that for all $u \in H^m(\Omega)$

$$(1.5) \quad C_1 \inf_{\chi\in\mathcal{P}_{m-1}(\Omega)} \|u-\chi\|_{m,\Omega} \le |u|_{m,\Omega} \le C_2 \inf_{\chi\in\mathcal{P}_{m-1}(\Omega)} \|u-\chi\|_{m,\Omega}.$$

<u>Proof</u>: For any $u \in H^m(\Omega)$ and $\chi \in \mathcal{P}_{m-1}(\Omega)$

$$|u|_{m,\Omega} \equiv |u-\chi|_{m,\Omega} \le \|u-\chi\|_{m,\Omega},$$

which proves the right hand inequality of (1.5). In order to prove the left hand inequality we use (1.4). From (1.4), for any $\chi \in \mathcal{P}_{m-1}(\Omega)$

$$\|u-\chi\|_{m,\Omega}^2 \le |u|_{m,\Omega}^2 + \sum_{|\beta|\le m-1} \left[\int D^\beta (u-\chi) dx\right]^2.$$

We shall show by induction that χ may be chosen so that each term of the sum over $|\beta| \leq m-1$ vanishes, i.e.,

$$\int_\Omega D^\beta(u-\chi)dx = 0. \tag{1.6}$$

When $m = 1$, then χ is a constant, chosen to be

$$\chi = \frac{\int_\Omega udx}{\int_\Omega dx} .$$

Assume now that this can be done for $|\beta| \leq m-2$. Any $\chi \in \mathcal{P}_{m-1}(\Omega)$ has the form $\chi = \sum_{|\beta|\leq m-1} \frac{C_\beta}{\beta!} x^\beta$, where the C_β are constants. Hence for any derivative $D^\beta\chi$ of order $|\beta| = m-1$, $D^\beta\chi = C_\beta$ and we choose

$$C_\beta = \frac{\int_\Omega D^\beta udx}{\int_\Omega dx} .$$

Now $\chi = \chi_{m-1} + \chi_{m-2}$ where $\chi_{m-2} = \sum_{|\beta|\leq m-2} \frac{C_\beta}{\beta!} x^\beta$ and χ_{m-1} is known. χ_{m-2} is now determined by applying the induction hypothesis to $u-\chi_{m-1}$, which completes the proof.

Another useful Poincaré inequality is as follows:

<u>Lemma 1.4</u>. Let Ω be a bounded open set in $\mathbb{R}^2$, $\Omega \subset D$, where $D = \{(x,y):0 < x_1 < d,\ 0 < x_2 < d\}$. Then if $u \in H_0^m(\Omega)$

(1.7) $$\|u\|_\Omega \le d^m |u|_{m,\Omega}.$$

<u>Proof</u>: We shall give the proof in the case that $m = 1$, the general case is left to the reader. By density, it suffices to consider $u \in C_0^1(\bar{\Omega})$ and extend u by 0 to D. Then since $u(0,x_2) = 0$

$$u(x_1,x_2) = \int_0^{x_1} \frac{\partial u}{\partial \zeta}(\zeta,x_2)d\zeta.$$

Squaring and using the Schwarz inequality

$$u^2(x_1,x_2) \le d \int_0^d (\frac{\partial u}{\partial \zeta}(\zeta,x_2))^2 \, d\zeta.$$

The result now easily follows after integrating both sides with respect to x_1 and x_2.

<u>Remark 1.2</u>. An immediate consequence of Lemma 1.4 is that for $u \in H_0^1(\Omega)$, where Ω is an open bounded set with $\text{diam}(\Omega) = d$,

(1.8) $$\|\nabla u\|_\Omega^2 \le \|u\|_{1,\Omega}^2 \le (1+d^2)\|\nabla u\|_\Omega^2 .$$

Hence $(\nabla u,\nabla v)_\Omega$ is an inner product on $H_0^1(\Omega)$ which induces a norm equivalent to the $H^1(\Omega)$ norm.

1C. Two Sobolev Inequalities.

We shall now state some special Sobolev inequalities.

Lemma 1.5. Let Ω be a bounded domain in $\mathbb{R}^2$ which is either polygonal or has a C^1 boundary. There exists a constant $C > 0$, depending only on Ω such that for all $u \in W^{2,1}(\Omega)$

(1.9) $$\|u\|_{C(\bar{\Omega})} \leq C\|u\|_{W^{2,1}(\Omega)}$$

and

(1.10) $$\|u\|_{1,\Omega} \leq C\|u\|_{W^{2,1}(\Omega)}.$$

Here

$$\|u\|_{W^{2,1}(\Omega)} = \sum_{|\beta|\leq 2} \int_\Omega |D^\beta u|dx.$$

We shall also need the semi-norm

$$|u|_{W^{2,1}(\Omega)} = \sum_{|\beta|=2} \int_\Omega |D^\beta u|dx.$$

For a proof of Lemma 1.5 we refer the reader to Adams [1975].

2. Some Finite Element Spaces and Their Properties.

Unless otherwise stated, in this section Ω will be a polygonal domain in $\mathbb{R}^2$. Our aim here is to define certain triangulations of Ω and discuss some basic properties of special finite dimensional subspaces of $H^1(\Omega)$. In particular, we shall be mainly concerned with a special but representative example of so called finite element spaces, the continuous piecewise linear functions. We shall discuss some approximation properties, inverse properties and superapproximation properties. A brief discussion of some other (higher order piecewise polynomial) finite element spaces and the case when Ω is not polygonal will be given at the end of this section. These functions will be used to approximate the solutions of elliptic boundary value problems in the subsequent sections.

2A. Quasi-uniform Families of Triangulations.

Let $0 < h \leq 1$ be a discretization parameter, and for each such h, let $\{\tau\}^h$ denote a partition of Ω into disjoint triangles τ_j, $j = 1,\ldots,N(n)$ such that:

i) $\bar{\Omega} = \bigcup_{j=1}^{N(n)} \bar{\tau}_j$.

ii) The common edges of any two adjoining triangles coincide.

iii) The maximum length of a side of any triangle τ_j is $\leq h$.

iv) The triangulations are quasi-uniform (of size h), i.e., the area of any triangle $\tau_j \in \{\tau\}^h$ is bounded below by ch^2, where $c > 0$ is independent of h and τ_j.

Remark 2.1. Property iv) may be restated in an equivalent form as follows: There exists a constant $m > 0$, independent of h, such that every triangle τ_j is contained in a ball of radius h and contains a ball of radius mh. Roughly speaking the triangles in $\{\tau\}^h$ are of the "same" size and do not degenerate in that the minimum angle of any triangle is uniformly bounded away from zero independent of h.

2B. Finite Element Spaces.

We start with the construction of special finite element spaces which are subspaces of $H^1(\Omega)$ and $H_0^1(\Omega)$ respectively. Let $S^h(\Omega)$ denote the continuous functions on $\overline{\Omega}$ which are linear on each triangle of $\{\tau\}^h$, and let $S_0^h(\Omega)$ be the subspace of $S^h(\Omega)$ consisting of those functions which vanish on $\partial\Omega$. The triangles in $\{\tau\}^h$ are called "elements" and in general any member of the space of functions which are piecewise polynomials (relative to the triangulation $\{\tau\}^h$) is called a finite element. The particular finite element space we have considered here, the space of continuous piecewise linear functions, was first suggested for approximating solutions of partial differential equations by R. Courant in 1942.

Obviously, a linear function on a triangle is uniquely determined by specifying its values at the vertices. These vertices shall be called nodes and we assume that they are ordered. Every $\varphi \in S^h(\Omega)$ may be uniquely written in the form

$$\varphi(x) = \sum \alpha_i \varphi_i(x), \tag{2.1}$$

where $\varphi_i \in S^h(\Omega)$ is that basis function which is 1 at the ith "node" and zero at all other nodes (see Figure 2.1).

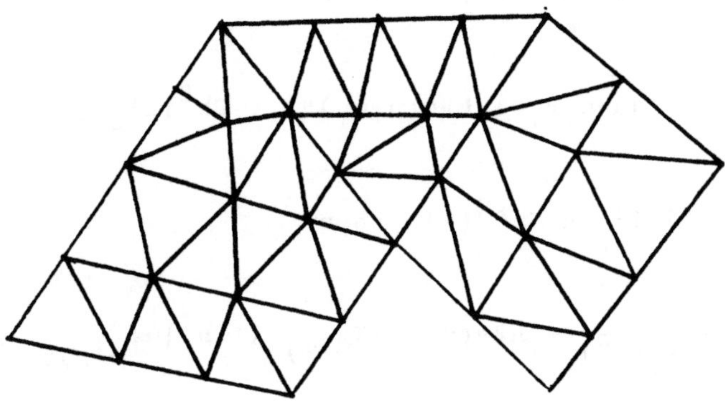

Figure 2.1

For $S_0^h(\Omega)$, we only need those φ_i which are 1 at those nodes and which are interior points in Ω.

Since these functions are to be used to approximate the solutions of boundary value problems, we will first discuss some of their approximation properties. In order to do so, let us first introduce the notion of the interpolant of a function.

2C. Some Approximation Properties of $S^h(\Omega)$ and $S_0^h(\Omega)$.

Definition. For $u \in C(\bar{\Omega})$, the interpolant $u_I \in S^h(\Omega)$ of u is defined to be the unique continuous piecewise linear function which is equal to u at the nodal points of the triangulation $\{\tau\}^h$.

We shall use a method due to Bramble and Hilbert [1970], [1971] to prove

Theorem 2.1. Let τ be any triangle in $\{\tau\}^h$.

i) If $u \in H^2(\tau)$ then

$$\|u-u_I\|_\tau + h\|\nabla(u-u_I)\|_\tau \leq Ch^2|u|_{2,\tau}. \tag{2.2}$$

ii) If $u \in C^2(\bar{\tau})$ then

$$\|u-u_I\|_{C(\bar{\tau})} + h\|\nabla(u-u_I)\|_{C(\bar{\tau})} \leq Ch^j|u|_{C^j(\bar{\tau})}, \quad j = 1,2, \tag{2.3}$$

where C is independent of u, h and τ.

An immediate consequence of Theorem 2.1 is the following:

Corollary 2.1. Let $\bar{D}_h$ be a mesh domain, i.e., the union of the closures of any set of triangles in $\{\tau\}^h$.

i) If $u \in H^2(D_h)$ (resp. $H_0^1(D_h) \cap H^2(D_h)$), then $u_I \in S^h(D_h)$ (resp. $S_0^h(D_h)$) satisfies

$$\|u-u_I\|_{D_h} + h\|\nabla(u-u_I)\|_{D_h} \leq Ch^2|u|_{2,D_h}. \tag{2.4}$$

ii) If $u \in C^2(\bar{D}_h)$ (resp. $C^2(\bar{D}_h) \cap C_0(\bar{D}_h)$), then $u_I \in S^h(D_h)$ (resp. $S_0^h(D_h)$) satisfies

$$\|u-u_I\|_{C(\bar{D}_h)} + h\|\nabla(u-u_I)\|_{L^\infty(\bar{D}_h)} \leq Ch^j|u|_{C^j(\bar{D}_h)}, \quad j = 1,2, \tag{2.5}$$

where C is independent of u, h and D_h.

The proof of (2.5) follows easily from (2.3). The inequality (2.4) follows from (2.2) by squaring both sides, summing over all triangles τ in D_h and then taking square roots.

Remark 2.2. (Important) There is a standard first step which is often used in proving various properties, e.g. approximation properties, inverse properties (see Theorem 2.2), etc. of the subspaces $S^h(\Omega)$. This consists of mapping each of the triangles $\tau \in \{\tau\}^h$ onto a single reference triangle τ^*, with vertices say (0,0), (1,0), (0,1), by means of an orientation preserving affine mapping. Let y_1, y_2 refer to the new coordinates on τ^* where

$$y_1 = \alpha_1 + \beta_1 x_1 + \gamma_1 x_2$$

$$y_2 = \alpha_2 + \beta_2 x_2 + \gamma_2 x_2 .$$

for some constants α_1, β_1, γ_1, α_2, β_2, γ_2, which are completely determined by the assignment of the vertices of τ to the vertices of τ^*. It is important to note that a linear function on τ is transformed into a linear function on τ^*. In fact more generally, a polynomial of degree $\leq r$ is transformed into polynomial of degree $\leq r$. A desired result is then hopefully proved on τ^* (a unit size domain) and the result transformed back to τ. The mapping from any such τ to τ^* may be thought of as a composition of two affine mappings. The first is a mapping onto a "reference" triangle τ_h^* of size h, i.e., with vertices say $(0,0)$, $(h,0)$, $(0,h)$, and then a simple uniform scaling to τ^*. Because of our assumption of quasi-uniformity, the mapping from $\tau \to \tau_h^*$ and its inverse are "nice" with Jacobians uniformly bounded away from zero, independent of τ and h. Under these transformations for example

$$C_1 \|u\|_{m,\tau} \leq \|u\|_{m,\tau_h^*} \leq C_2 \|u\|_{m,\tau} ,$$

with C_1 and C_2 independent of h and τ. Hence in what follows we shall assume without loss of generality that $\tau = \tau_h^*$.

<u>Remark 2.4</u>. The interpolation operator is the identity on $S^h(\Omega)$, i.e., if $\chi \in S^h(\Omega)$ then $\chi_I = \chi$.

<u>Proof of Theorem 2.1</u>: In view of Remark 2.3 we will prove the result with τ replaced by τ_h^* the "reference" triangle of size h. Transform τ_h^* to τ^* by the scaling

$$y_1 = \frac{x_1}{h} , \; y_2 = \frac{x_2}{h} , \tag{2.6}$$

and let $u^*(y_1,y_2) = u(hy_1,hy_2)$ and $u_I^*(y_1,y_2) = u_I(hy_1,hy_2)$. Now

$$\begin{aligned} (2.7) \qquad & u_I^*(y_1,y_2) \\ & = u^*(0,0)+(u^*(1,0)-u^*(0,0))y_1+(u^*(0,1)-u^*(0,0))y_2 . \end{aligned}$$

From this it easily follows (using a crude bound with the triangle inequality) that

$$\|u_I^*\|_{\tau^*} \le 5\|u^*\|_{C(\overline{\tau^*})}. \tag{2.8}$$

and

$$\|\nabla u_I^*\|_{\tau^*} \le \|u_I^*\|_{C(\tau^*)} \le 5\|u^*\|_{C(\overline{\tau^*})}. \tag{2.9}$$

Let us prove (2.2). The proof of (2.3) will be left to the reader. Since $u \in H^2(\tau^*)$, we have from (2.8) and the Sobolev Inequality (1.9) that

$$\|u^*-u_I^*\|_{\tau^*} \le \|u^*\|_{\tau^*} + \|u_I^*\|_{\tau^*} \le C\|u^*\|_{W^{2,1}(\tau^*)} \le C\|u^*\|_{2,\tau^*}.$$

Here C is independent of u^* and h. Since the linear interpolant reproduces linear functions, it follows that $(u^*-\chi^*)_I = u_I^*-\chi^*$, where χ^* is any linear function of the form

$$\chi^* = \alpha + \beta y_1 + \gamma y_2.$$

Hence

$$\begin{aligned} \|u^*-u_I^*\|_{\tau^*} &\le \|(u^*-\chi^*)-(u^*-\chi^*)_I\|_{\tau^*} \\ &\le C\|u-\chi^*\|_{2,\tau^*} \\ &\le C \inf_{\chi^*\in\mathcal{P}_1(\tau^*)} \|u-\chi^*\|_{2,\tau^*} . \end{aligned} \tag{2.10}$$

Similarly one derives

$$(2.11) \qquad \|\nabla(u^*-u_I^*\|_{\tau^*} \le C \inf_{\chi^*\in\mathcal{P}_1(\tau^*)} \|u^*-\chi^*\|_{2,\tau^*}.$$

From Lemma 1.3 we have that

$$(2.12) \qquad \inf_{\chi^*\in\mathcal{P}_1(\tau^*)} \|u^*-\chi^*\|_{2,\tau^*} \le C|u^*|_{2,\tau^*},$$

where we emphasize that the term on the right in (2.12) involves only the semi-norm of second derivatives. Thus

$$(2.13) \qquad \|u^*-u_I^*\|_{\tau^*} \le C|u^*|_{2,\tau^*},$$

and

$$(2.14) \qquad \|\nabla(u^*-u_I^*)\|_{\tau^*} \le C|u^*|_{2,\tau^*}.$$

The proof of (2.2) will follow from (2.13) and (2.14) by scaling these inequalities back to $\tau = \tau_h^*$. Using the transformation $y_1 = \frac{x_1}{h}$, $y_2 = \frac{x_2}{h}$, it follows that

$$\int_{\tau^*} (u^*-u_I^*)^2 dy = h^{-2} \int_{\tau_h^*} (u-u_I)^2 dx.$$

Hence

$$|u^*|^2_{2,\tau^*} \equiv \int_\tau \left[\frac{\partial^2 u^*}{\partial y_1^2}\right]^2 + \left[\frac{\partial^2 u^*}{\partial y_1 \partial y_2}\right]^2 + \left[\frac{\partial^2 u^*}{\partial y_2^2}\right]^2 dy$$

$$= h^2 \int_{\tau_h^*} \left[\frac{\partial^2 u}{\partial x_1^2}\right]^2 + \left[\frac{\partial^2 u}{\partial x_1 \partial x_2}\right]^2 + \left[\frac{\partial^2 u}{\partial x_2^2}\right]^2 dx$$

$$= h^2 |u|^2_{2,\tau_h^*} .$$

Similarly

$$\|\nabla(u^* - u_I^*)\|_{\tau^*} = \|\nabla(u - u_I)\|_{\tau_h} .$$

Combining these last expressions with (2.13) and (2.14) gives (2.2) which completes the proof.

2D. Inverse Properties of $S^h(\Omega)$.

For an excellent and detailed exposition of approximation properties of finite elements we refer the reader to Ciarlet [1978].

We now turn to proving some other fundamental properties of the subspaces $S^h(\Omega)$. The first results belong to a class of useful inequalities usually referred to as inverse properties. We shall consider only those which will be needed in what follows.

Theorem 2.2. Let $\{\tau\}^h$ be a quasi-uniform family of triangulations and $S^h(\Omega)$ be as above. There exists a constant

C such that if $\overline{D}_h$ is the closure of the union of any set of triangles in $\{\tau\}^h$, then for any $\chi \in S^h(\Omega)$

$$\|\chi\|_{1,D_h} \le Ch^{-1}\|\chi\|_{D_h} \tag{2.15}$$

and

$$\|\chi\|_{C(\overline{D}_h)} \le Ch^{-1}\|\chi\|_{D_h}. \tag{2.16}$$

Here C is independent of h, χ and D_h.

Remark 2.4. These inequalities say that on the finite dimensional subspaces S^h, we may bounded "stronger norms" by "weaker norms", with constants which depend on h. The analogous inequalities, of course, do not hold for example for if we replace $\chi \in S^h(\Omega)$ with $\chi \in H^1(D_h)$ or $C(\overline{D}_h)$, respectively. Notice that as $h \to 0$ the above estimates degenerate in a very specific way.

Proof of Theorem 2.2. We look at any triangle $\tau \in \{\tau\}^h$. In view of Remark 2.2, we shall again assume for simplicity that $\tau = \tau_h^*$. As in the proof of Theorem 2.1, τ_h^* is transformed to τ^* via $y_1 = \frac{x_1}{h}$, $y_2 = \frac{x_2}{h}$. Let χ be the transformed χ^*. Note, again, that the space of linear functions on τ_h^* are mapped onto the space of linear functions on τ^*, which is a fixed finite dimensional space (of dimension 3 in our special case) independent of h. Since all norms on a fixed finite dimensional space are equivalent, we have, for example, that for any linear χ^* on τ^*

$$\|\chi^*\|_{1,\tau^*} \le C\|\chi^*\|_{\tau^*} \tag{2.17}$$

and

$$\|\chi^*\|_{C(\overline{\tau}^*)} \le C\|\chi^*\|_{\tau^*} . \tag{2.18}$$

In (2.17) and (2.18) the constant C is independent of h.

The result (2.15) follows easily by transforming (2.17) back to τ_h and then τ, squaring both sides and summing over all $\tau \in D_h$. The inequality (2.16) follows by transforming (2.18) back to τ_h. We leave the details to the reader.

2E. Super-approximation Property.

The next result that will be proved is known as the super-approximation property. The approximation result (2.4) deals with how well the subspace $S^h(\Omega)$ can approximate a general function in $H^2(\Omega)$. We now consider the question of how well the special functions of the form $\omega(x)\chi(x)$ can be approximated, where in applications $\omega(x)$ is usually a fixed function and $\chi(x)$ is any function in $S^h(\Omega)$. Functions of this form arise very often in the analysis of the finite element method and the super-approximation property has proved to be an invaluable tool. The super-approximation property was first used in Nitsche and Schatz [1972], [1974].

Theorem 2.3. Let $\overline{D}_h$ be any mesh domain, i.e., the union of the closures of any set of triangles in $\{\tau\}^h$. Let $\omega(x) \in C^2(\overline{D}_h)$, then there exists a constant $C > 0$ such that for all $\chi \in S^h(D_h)$

$$(2.19) \qquad \|\omega\chi-(\omega\chi)_I\|_{D_h} + h\|\nabla(\omega\chi-(\omega\chi)_I)\|_{D_h}$$

$$\leq Ch^2(|\omega|_{C^2(\overline{D}_h)}\|\chi\|_{D_h} + |\omega|_{C^1(\overline{D}_h)}\|\nabla\chi\|_{D_h}).$$

Here C is independent of h.

Remark 2.5. If $\omega \equiv 1$ on $\overline{D}_h$, then the right hand side of (2.19) vanishes. The inequality (2.19) says that for products of functions of this type, the interpolant yields an order h^2 approximation in $L_2(D_h)$ and an order h approximation in $H_1(D_h)$, but in contrast to (2.4) the terms on the right only involve χ and its gradient in $L_2(D_h)$. It is for this reason that this is usually referred to as the super-approximation property.

Proof: Let τ be any triangle in D_h. Using the approximation result (2.2)

$$(2.20) \qquad \|\omega\chi-(\omega\chi)_I\|_\tau + h\|\nabla(\omega\chi-(\omega\chi)_I)\|_\tau \leq Ch^2|\omega\chi|_{2,\tau}.$$

Since $\dfrac{\partial^2\chi}{\partial x_i\partial x_j} = 0$ for linear χ, we have

$$\frac{\partial^2(\omega\chi)}{\partial x_i \partial x_j} = \left[\frac{\partial^2 \omega}{\partial x_i \partial x_j}\right]\chi + \frac{\partial\omega}{\partial x_i}\frac{\partial\chi}{\partial x_j} + \frac{\partial\omega}{\partial x_j}\frac{\partial\omega}{\partial x_i}.$$

Hence evaluating the semi-norm on the right of (2.20)

$$(2.21) \qquad |\omega\chi|_{2,\tau} \le C(|\omega|_{C^2(\bar{\tau})}\|\chi\|_\tau + |\omega|_{C^1(\bar{\tau})}\|\nabla\chi\|_\tau).$$

The desired result now follows from (2.20) and (2.21) by appropriately summing over triangles in D_h.

2F. A Sobolev Inequality on $S^h(\Omega)$.

We now consider one more technical result which will be used later on. A result due to Sobolev says that, in two dimensions, the maximum norm of a function defined on $\bar{\Omega}$ can be bounded by $C(S)\|u\|_{S,\Omega}$ for any $S > 1$, where $C(S)$ depends on S. Emotionally speaking, it can "almost" be bounded by the $H^1(\Omega)$ norm. However, if we restrict u to the subspaces $S_0^h(\Omega)$, then the next result says that this can be done with a constant which blows up as $h \to 0$ in a very specific way.

<u>Theorem 2.4</u>. Let Ω be a polygonal domain and $S_0^h(\Omega)$ as above. There exists a constant C such that for any $\chi \in S_0^h(\Omega)$

$$(2.22) \qquad \|\chi\|_{C(\bar{\Omega})} \le C(\ln 1/h)^{1/2}\|\nabla\chi\|_\Omega,$$

where C is independent of h and χ.

<u>Remark 2.6</u>. (2.22) still holds for $\chi \in S^h(\Omega)$ provided $\|\nabla\chi\|_\Omega$ is replaced by $\|\chi\|_{1,\Omega}$. We shall give a proof which depends on the fact that χ vanishes on $\partial\Omega$.

<u>Proof</u>: Let $x_0 \in \bar{\Omega}$ be such that $\|\chi\|_{C(\bar{\Omega})} = |\chi(x_0)|$. Extend χ as zero outside of $\bar{\Omega}$ to a ball $B(x_0,d)$ or radius d such that $\bar{\Omega} \subset B(x_0,d)$. The function $G(x_0,x) = \frac{1}{2\pi} \ln|\frac{x-x_0}{d}|$ is the Green's function on $B(x_0,d)$ for $-\Delta$ with singularity at x_0. Then

$$\chi(x_0) = \int_{B(x_0,d)} \nabla G(x_0 x)\nabla\chi(x)dx$$

$$= \int_{B(x_0,h)} \nabla G(x_0 x)\nabla\chi(x)dx + \int_{B(x_0,d)\setminus B(x_0,h)} \nabla G(x_0,x)\nabla\chi(x)dx$$

$$= I_1 + I_2.$$

Let $\rho = |x-x_0|$, then $|\frac{\partial G}{\partial x_i}| \leq \frac{C}{\rho}$, $i = 1,2$. Using polar coordinates

$$\|\nabla G\|_{L^2(B(x_0,h))} \leq C \int_0^h d\rho = ch.$$

Hence using an inverse property on $S_0^h(\Omega)$,

$$|I_1| \leq \|\nabla G\|_{L^1(B(x_0,h))} \|\nabla\chi\|_{L^\infty(\Omega)} \leq Ch^{-1}\|\nabla G\|_{L^1(B(x_0,h))} \|\nabla\chi\|_\Omega$$

$$\leq \|\nabla\chi\|_\Omega .$$

Using the Cauchy-Schwarz inequality

$$|I_2| \le \|\nabla\chi\|_\Omega \|\nabla G\|_{B(x_0,d)/B(x_0,h)}$$

$$\le C\|\nabla\chi\|_\Omega \left[\int_h^d \frac{1}{\rho}\, d\rho\right]^{1/2}$$

$$\le C(\ell n\ \frac{1}{h})^{1/2} \|\nabla\chi\|_\Omega .$$

Taken together these inequalities prove the result.

Let us remark that this result still holds if $\chi \in S^h(\Omega)$ provided $\|\nabla\chi\|_\Omega$ is replaced by $\|\chi\|_{1,\Omega}$ on the right hand side.

2G. Some Generalizations to Higher Order Elements and Non-Polygonal domains.

Let us end this section by considering some other possible choices of approximating functions and then the case where Ω is not a polygonal. We have discussed the situation where our approximating functions are continuous piecewise linear functions on Ω, a polygonal domain. For this space, the interpolant is second order accurate (in h) in $L_2(\Omega)$ and $C(\overline{\Omega})$ provided $u \in H^2(\Omega)$ or $C^2(\overline{\Omega})$, respectively. This is the best that one can expect in general. Higher order accuracy can be obtained by using functions which are higher order polynomials on each triangle provided u is smoother than above. For example, let $S^{h,r}(\Omega)$ denote the space of functions which are continuous on $\bar{\Omega}$ and which are polynomials of degree r-1, where $r \geq 2$, on each triangle $\tau \in \{\tau\}^h$. These are uniquely determined by specifying their values at nodal points. Rather than giving a general discussion of what these are let us illustrate them in two important cases. Notice in these cases there are nodal points which are not the vertices of triangles.

Quadratic polynomial in τ Cubic polynomial on τ

• = nodal points

Figure 2.2

For $S^{h,3}(\Omega)$, we take the vertices of τ and the midpoints of each side as nodes, and for $S^{h,4}(\Omega)$, these are the vertices of τ, two equally spaced points on each side of τ and one interior point. It is obvious then how to define the interpolant, and following the proof of Theorem 2.1, one concludes that if $u \in H^j(\Omega)$ and $C^j(\bar{\Omega})$ respectively, then

$$\|u-u_I\|_\Omega + h\|\nabla(u-u_I)\|_\Omega \leq Ch^j|u|_{j,\Omega}, \quad j = 2,\dots,r, \tag{2.23}$$

and

$$\|u-u_I\|_{C(\bar{\Omega})} + h\|\nabla(u-u_I)\|_{C(\bar{\Omega})} \leq Ch^j|u|_{C^j(\bar{\Omega})}, \quad j = 1,\dots,r. \tag{2.24}$$

Thus for example, the error in the interpolant is of order h^r in $L^2(\Omega)$ and $C(\overline{\Omega})$, respectively, provided $u \in H^r(\Omega)$ and $C^r(\overline{\Omega})$, respectively. There is a large difference between the accuracy obtained with a piecewise linear function, $O(h^2)$, and say a piecewise cubic $O(h^4)$. The "higher order" subspaces are very efficient in approximating the solutions of boundary value problems which are sufficiently smooth, even though they are more complicated to implement and require more computational effort.

Let us now consider a case where Ω is not a polygonal domain. For simplicity let Ω be a convex domain with, say, a C^∞ boundary. In this case we triangulate Ω as before. Let $\overline{\Omega}_h = \cup \overline{T}_i$ where we further require that the vertices which lie on $\partial\Omega_h$ also lie on $\partial\Omega$ (see Figure 2.3).

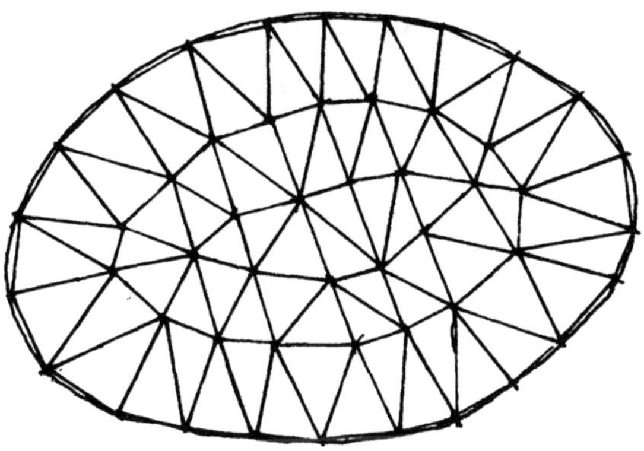

Figure 2.3

Under these conditions $\Omega_h \subset \Omega$ and because we have triangulated with triangles with straight edges, roughly of size h, we are not able to fit the boundary exactly. However it is easily seen that

$$(2.25) \qquad \operatorname{dist}(\partial\Omega_h, \partial\Omega) \le Ch^2.$$

Suppose we wished to approximate functions which vanish on $\partial\Omega$. We may take $S_0^h(\Omega)$ to be, for example, the continuous piecewise linear functions on $\bar{\Omega}_h$ which vanish on $\partial\Omega_h$ and simply extend them by zero outside of $\bar{\Omega}_h$. It is not hard to see that the estimates (2.23) and (2.24) still hold (in the case $r = 2$). For higher order piecewise polynomial spaces there is difficulty in that (2.25) is not enough. This difficulty can be overcome by using "triangles" with curved sides which better fit the boundary and which are mapped onto a triangle with straight sides where the subspaces are taken to be the higher order polynomials. These so called isoparametric elements are discussed at length in Ciarlet [1978], to which we refer the reader.

3. Orthogonal Projections into Finite Element Spaces in $L_2(\Omega)$, $H^1(\Omega)$ and $H_0^1(\Omega)$.

In this section we shall begin our study of finite element methods by considering the properties of the orthogonal projection into finite element subspaces of $L_2(\Omega)$, $H^1(\Omega)$ and $H_0^1(\Omega)$ and finally prove basic error (energy) estimates in these norms. The use of finite elements will be motivated from the points of view of ease of computation, adaptability to different geometries and accuracy. We shall end this section with a discussion of the properties of the L_2 projection in other norms, namely H^1 and negative norm Sobolev spaces. The techniques used to prove these estimates are useful in other situations. The connection between the H^1 and H_0^1 projections and various elliptic boundary value problems will be discussed in Section 4B.

3A. Orthogonal Projections in L_2, H^1 and H_0^1. Basic Error Estimates.

Let H be a real Hilbert space with inner product $(\cdot,\cdot)_H$ and norm $\|\cdot\|_H$. Let S^N be a finite dimensional subspace with basis $\varphi_1,\ldots,\varphi_N$. Then the following is well known:

<u>Lemma 3.1</u>. For each $u \in H$, there exists a unique $u_N \in S^N$ satisfying

$$(3.1) \qquad (u-u_N,\varphi)_H = 0 \quad \text{for all} \quad \varphi \in S^N.$$

u_N is the orthogonal projection of u into S^N. Equivalently u_N is the best approximation in S^N to u in the norm of H, i.e.,

$$(3.2) \qquad \|u-u_N\|_H = \inf_{\chi\in S^N} \|u-\chi\|_H .$$

<u>Remark 3.1</u>. In order to compute u_N from (3.1) we set

$$(3.3) \qquad u_N = \sum_{i=1}^{N} a_i\varphi_i .$$

Choosing the test functions $\varphi = \varphi_j$, $j = 1,\dots,N$, in (3.4) leads to a linear system of equations for the unknowns a_i

$$(3.4) \qquad \sum_{i=1}^{N} a_i(\varphi_i,\varphi_j)_H = (u,\varphi_j)_H,\ j = 1,\dots,N.$$

The matrix $A_{ij} = (\varphi_i,\varphi_j)_H$ is the Gramm matrix of the basis functions $\varphi_1,\dots,\varphi_N$ in the inner product in H. In practice, it is this matrix which must be computed and inverted. It is important that both of these operations not be too costly.

<u>Example 1</u>. <u>The L_2 projection P_h</u>. We shall begin our study of finite element methods by considering a simple projection method, namely the L_2 projection. Let $H = L_2(\Omega)$, where for simplicity we take Ω to be a polygonal domain. Let $u \in L_2(\Omega)$. We seek the best approximation to u in some

subspace S^N. There are many possible choices of S^N which are reasonable and used in practice. Some examples are:

i) Polynomials of degree $\leq K$, i.e., $\mathcal{P}_K(\Omega)$.

ii) Finite trigonometric series.

iii) Finite element subspaces, for example $S^h(\Omega)$.

Let us note that in general for polygonal domains the matrix A_{ij} is full in the cases i) and ii) and in these cases to compute each A_{ij} requires an integration over the whole domain (there are exceptions of course such as the trigonometric functions on a rectangle). In contrast the choice iii) of $S^N = S^h(\Omega)$ leads to more easily computable sparse matrices. This is easy to see since the only possible non-zero entries are those for which $(\varphi_i, \varphi_j)_\Omega$ corresponds to basis functions of nodes i and j which are connected by the edge of a triangle. The integral $(\varphi_i, \varphi_j)_\Omega$ extends over a region of roughly area h^2 and involves integrating polynomials on triangles. As we shall see later on, $S^h(\Omega)$ has additional convenient properties which are also shared by other finite element spaces.

From now on we shall make the choice $S^N = S^h(\Omega)$ and set

$$(3.5) \qquad u_N = P_h u,$$

where $P_h u$ is the L_2 projection of u into S^h. Thus $P_h u$ is defined by

$$(3.5) \qquad (u - P_h u, \varphi)_\Omega = 0 \quad \text{for all} \quad \varphi \in S^h(\Omega)$$

or equivalently

$$\|u-P_h u\| = \inf_{\chi\in S^h(\Omega)} \|u-\chi\|_\Omega. \tag{3.6}$$

<u>Example 2. The orthogonal projection in $H^1(\Omega)$, R_h^N.</u> Here we take $H = H^1(\Omega)$ with the usual inner product

$$(u,\varphi)_{1,\Omega} = (\nabla u,\nabla\varphi)_\Omega + (u,\varphi)_\Omega. \tag{3.7}$$

For S^N we choose $S^N = S^h(\Omega)$, and for u_N, the best approximation in $S^h(\Omega)$ to u, we define

$$u_N = R_h^N u. \tag{3.8}$$

$R_h^N u$ satisfies

$$(\nabla(u-R_h^N u),\nabla\varphi) + (u-R_h^N u,\varphi) = 0 \quad \text{for all} \quad \varphi \in S^h(\Omega). \tag{3.9}$$

Furthermore

$$\|u-R_h^N u\|_{1,\Omega} = \inf_{\chi\in S^h(\Omega)} \|u-\chi\|_{1,\Omega}. \tag{3.10}$$

Note that, as for the L_2 projection the matrix $A_{ij} = (\varphi_i,\varphi_j)_{1,\Omega}$ is sparse (and called the stiffness matrix by engineers).

Example 3. An orthogonal projection in $H_0^1(\Omega), R_h$. Here we take $H = H_0^1(\Omega)$ with the inner product (see Remark 1.2).

(3.11) $$(u,\varphi)_{1,\Omega} = (\nabla u, \nabla\varphi)_\Omega.$$

For S^N we choose $S^N = S_0^h(\Omega)$ and for u_N, the best approximation in $S_0^h(\Omega)$ to u, we define

(3.12) $$u_N = R_h u.$$

$R_h u$ satisfies

(3.13) $$(\nabla(u-R_h u), \nabla\varphi) = 0 \quad \text{for all} \quad \varphi \in S_0^h(\Omega).$$

Equivalently,

(3.14) $$\|\nabla(u-R_h u)\|_\Omega = \inf_{\chi\in S_0^h(\Omega)} \|\nabla(u-\chi)\|_\Omega.$$

An immediate consequence of the best approximation properties of these projections and the properties of the interpolant is the following basic error estimates.

Theorem 3.1. i) Let $u \in H^2(\Omega)$, then

(3.15) $$\|u-P_h u\|_\Omega \leq ch^2|u|_{2,\Omega}.$$

ii) Let $u \in H^2(\Omega)$, then

(3.16) $$\|(u-R_h^N u)\|_{1,\Omega} \leq ch|u|_{2,\Omega}.$$

iii) Let $u \in H_0^1(\Omega) \cap H^2(\Omega)$, then

(3.17) $$\|\nabla(u-R_h u)\|_{\Omega} \leq ch|u|_{2,\Omega}.$$

<u>Proof</u>: The inequalities (3.15), (3.16) and (3.17), respectively follow from (2.4) and (3.6), (3.10) and (3.14), respectively.

<u>3B. Further Estimates for the L_2 Projection, the Error in H^1 and Negative Norms, Duality</u>.

Theorem 3.1 gives estimates for the error in the norms of the spaces in which the various projections were taken. We may now ask what the error of these projections are in other norms. For example, what is the error of the L_2 projection in H^1? We shall first show in a very simple manner that the order of accuracy of this error in H^1 is one order less than that in L_2, which is analogous to the interpolant. A crucial assumption is that the mesh is quasi-uniform and hence inverse properties hold on the subspace. We shall also investigate how the error behaves in "weaker norms", in particular, negative order Sobolev norms. The proof here serves as an introduction to the more complicated Aubin-Nitsche duality method of estimating the error of the H^1 and H_0^1 projections in "weaker norms" namely $L_2(\Omega)$ and negative order Sobolev norms. The latter estimates will be discussed in Section 4. We return now to the L_2 projection.

We now look at the error in $H^1(\Omega)$ for $u-P_h u$. We obtain this estimate from the basic estimate (3.15) using a simple but useful device involving inverse properties.

Theorem 3.2. Let $u \in H^2(\Omega)$, then with $S^h(\Omega)$ as in Section 1

(3.18) $$\|\nabla(u-P_h u)\|_\Omega \leq ch|u|_{2,\Omega}.$$

Proof: Using the triangle inequality

$$\|\nabla(u-P_h u)\|_\Omega \leq \|\nabla(u-u_I)\|_\Omega + \|\nabla(u_I-P_h u)\|_\Omega,$$

where u_I is given in Section 2C. Since $u_I-P_h u \in S^h(\Omega)$, the inverse estimate (2.15) yields

$$\|\nabla(u_I-P_h u)\|_\Omega \leq ch^{-1}\|u_I-P_h u\|_\Omega$$

$$\leq ch^{-1}(\|u-P_h u\|_\Omega + \|u-u_I\|_\Omega),$$

where we have again used the triangle inequality. Collecting these inequalities it follows that

$$\|\nabla(u-P_h u)\|_\Omega \leq c(h^{-1}\|u-u_I\| + \|\nabla(u-u_I)\|)_\Omega + ch^{-1}\|u-P_h u\|_\Omega.$$

The inequality (3.18) now follows from this, (2.4) and (3.14).

3C. Negative Norm Errors for $u-P_hu$. Duality. Higher Order Elements.

We shall now examine the errors of the L_2 projection in negative Sobolev norms. These are defined as follows.

Definition. For m an integer, $m > 0$, let $H^{-m}(\Omega)$ be the completion of $C_0^\infty(\Omega)$ with respect to the norm

$$\|u\|_{-m,\Omega} = \sup_{\varphi \in H_0^m(\Omega)} \frac{(u,\varphi)}{\|\varphi\|_{m,\Omega}} .$$

Remark 3.2. $H^{-m}(\Omega) = (H_0^m(\Omega))'$, the dual space of $H_0^m(\Omega)$, where bounded linear functionals are defined relative to the L_2 inner product. Obviously

$$(u,\varphi)_\Omega \leq \|u\|_{-m,\Omega}\|\varphi\|_{m,\Omega},$$

for all $u \in H^{-m}(\Omega)$ and $\varphi \in H^m(\Omega)$. Note also that if $m_1 \geq m_2 \geq 0$, then

$$\|u\|_{-m_1,\Omega} \leq \|u\|_{-m_2,\Omega} \leq \|u\|_\Omega .$$

Let us now assume that we have triangulated Ω with a quasi-uniform triangulation and that, as in Section 2G, we have a finite element subspace $S^{h,r}(\Omega)$ which satisfies (2.23). That is

$$\text{(3.19)} \qquad \|u-u_I\|_\Omega + h\|\nabla(u-u_I)\|_\Omega \leq ch^j|u|_{j,\Omega}, \quad j = 2,\dots,r.$$

Using subspaces with these properties we will show

<u>Theorem 3.3</u>. Let $S^{h,r}(\Omega)$ be as in Section 2G and let $P_h u \in S^{h,r}(\Omega)$ satisfy

$$(u-P_h u, \varphi) = 0 \quad \text{for all} \quad \varphi \in S^{h,r}(\Omega).$$

Then for $-r \leq k \leq 1$, and $\max(k,0) \leq m \leq r$,

$$\|u-P_h u\|_{k,\Omega} \leq ch^{m-k} |u|_{m,\Omega}. \tag{3.20}$$

<u>Remark 3.3</u>. If $k = 0$ or 1, then (3.20) says that $P_h u$ has approximation properties in L_2 and H^1 similar to the interpolant u_I. However, it is different from the interpolant in that, for example, in the extreme case where $k = -r$

$$\|u-P_h u\|_{-r,\Omega} \leq ch^{2r} |u|_{r,\Omega}.$$

One can show that, in general, this does not hold with u_I replacing $P_h u$. The inequality (3.20) may be interpreted as follows: The error $u-P_h u$ is of order h^r from L_2 to H^r and by going "back" to lower negative norms say $H^{-r}(\Omega)$ its order of accuracy from $H^{-r}(\Omega)$ to $L_2(\Omega)$ is also h^r. Thus it is of order h^{2r} from $H^{-r}(\Omega)$ to $H^r(\Omega)$, double the order from L_2 to H^r (see Figure 3). We shall investigate an analogous, but more complicated, phenomena for the H^1 projections in Section 4. Let us end this remark by noting that one cannot in general prove an estimate of the form

$$\|u-P_hu\|_{-(r+1),\Omega} \le ch^{2r+1}|u|_{r,\Omega}.$$

Thus no improvements in accuracy for the L_2 projection can be found in general (using $S^{h,r}(\Omega)$) by going to norms with lower index than $H^{-r}(\Omega)$.

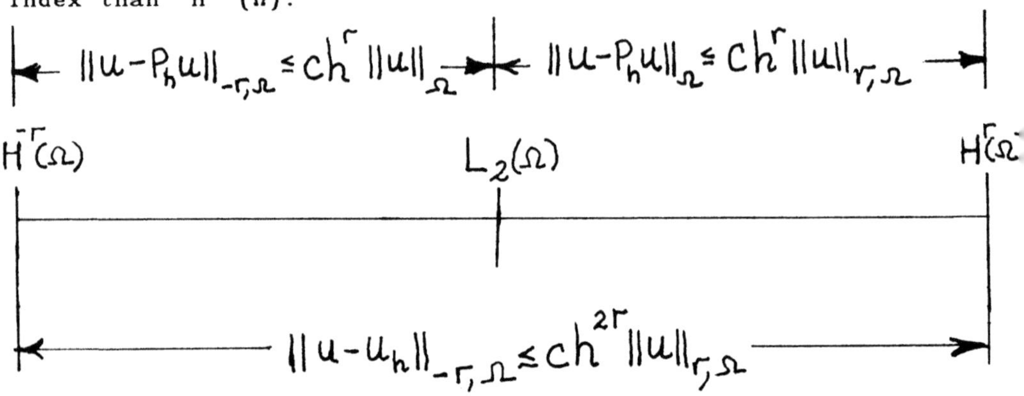

Figure 3.1

<u>Proof of Theorem 3.3</u>: For $k = 0$ or $k = 1$, (3.20) follows from (3.6) and (3.19). For simplicity we shall prove (3.20) in the extreme case $k = -r$. The proof for other values of k will be left to the reader. Now by definition of $H^{-r}(\Omega)$

$$\|u-P_hu\|_{-r,\Omega} = \sup_{v\in H_0^r(\Omega)} \frac{(u-P_hu,v)_\Omega}{\|v\|_{r,\Omega}}.$$

From the definition of P_hu and Schwarz's inequality we have

$$|(u-P_hu,v)_\Omega| = |(u-P_hu,v-P_hv)_\Omega| \le \|u-P_hu\|_\Omega\|v-P_hv\|_\Omega$$

$$\le ch^r\|v\|_{r,\Omega}\|u-P_hu\|_\Omega.$$

Hence

$$\|u-P_hu\|_{-r,\Omega} \le ch^r\|u-P_hu\|_\Omega.$$

The proof now follows from the case $k = 0$.

4. Galerkin Finite Element Methods for Second Order Elliptic Boundary Value Problems. Basic H^1 and L_2 Estimates.

In this section we shall introduce two model second order elliptic boundary value problems whose solutions will be approximated using the finite element method. These approximations will be identified with certain orthogonal projections discussed in Section 3. Basic error estimates in $H^1(\Omega)$ and $L_2(\Omega)$ will be discussed, the latter being obtained by an important duality argument due independently to Aubin [1967] and Nitsche [1968].

4A. Preliminaries. Two Model Elliptic Boundary Value Problems.

Here we will discuss two model elliptic boundary value problems to which the finite element method will be applied. In order to estimate the error between the approximate solution and the solution u of the boundary value problem, some information will be needed regarding the smoothness of u. This in general depends on the smoothness of the data and properties of the boundary $\partial\Omega$.

Let Ω be a bounded domain in $\mathbb{R}^2$ and for $f \in L_2(\Omega)$, let $u \in H_0^1(\Omega)$ be the weak solution of Dirichlet's Problem

$$\text{(4.1)} \qquad \begin{aligned} Lu &= -\Delta u = f \quad \text{in} \quad \Omega \\ u &= 0 \quad \text{on} \quad \partial\Omega. \end{aligned}$$

That is u is characterized in the following way:

Problem I (Dirichlet's Problem). For $f \in L^2(\Omega)$, find $u \in H_0^1(\Omega)$ satisfying

$$(4.2) \qquad (\nabla u, \nabla \varphi)_\Omega = (f, \varphi)_\Omega, \text{ for all } \varphi \in H_0^1(\Omega).$$

For our second model problem we consider the Neumann Problem

$$(4.3) \qquad \begin{aligned} Lu &= -\Delta u + u = f \quad \text{in } \Omega, \\ \frac{\partial u}{\partial n} &= 0 \quad \text{on } \partial\Omega. \end{aligned}$$

In weak form:

Problem II (Neumann's Problem). For $f \in L_2(\Omega)$, find $u \in H^1(\Omega)$ satisfying

$$(4.4) \qquad (\nabla u, \nabla \varphi)_\Omega + (u, \varphi)_\Omega = (f, \varphi)_\Omega \quad \text{for all } \varphi \in H^1(\Omega).$$

It is well known that both Problems I and II have unique solutions in $H_0^1(\Omega)$ and $H^1(\Omega)$, respectively. Since $f \in L_2(\Omega)$, it is reasonable to ask whether $u \in H^2(\Omega)$? The answer depends on the nature of $\partial\Omega$. We shall now state some well known regularity results for some classes of domains which will be considered in the following sections.

Theorem 4.1. Let $u \in H_0^1(\Omega)$ be the solution of (4.2) or $u \in H^1(\Omega)$ be the solution of (4.4).

i) If Ω has a C^k boundary $\partial\Omega$, $k \geq 2$ and $f \in H^{k-2}(\Omega)$, then $u \in H^k(\Omega)$ and

$$(4.5) \qquad \|u\|_{k,\Omega} \leq C\|f\|_{k-2,\Omega}.$$

ii) If Ω is a convex domain (for example a convex polygonal domain) and $f \in L^2(\Omega)$, then $u \in H^2(\Omega)$

$$(4.6) \qquad \|u\|_{2,\Omega} \leq C\|f\|_{\Omega}.$$

iii) Suppose Ω is a non-convex polygonal domain with interior angles $0 < \alpha_1 \leq \alpha_2 \leq \ldots \leq \alpha_{m-1} < \pi$, and $\pi < \alpha_m \leq 2\pi$, with the vertex of the non-convex corner with angle α_m at the origin. If $f \in L_2(\Omega)$, then for any fixed real number $t > 2\,(1-\pi/\alpha_m)$

$$(4.6a) \qquad \|u\|_{1,\Omega} + |u|_{2,\Omega,t} \leq C_t\|f\|_{\Omega}.$$

Here $|\cdot|_{2,\Omega,\rho,t}$ is the semi-norm defined by the weighted L_2 norms

$$|u|^2_{2,\Omega,t} = \sum_{|\beta|=2} \int_\Omega \rho^t (D^\beta u)^2 dx,$$

where $|\rho| = \sqrt{x_1^2+x_2^2}$, the distance to the origin.

<u>Remark 4.1</u>. Let us briefly discuss the results of Theorem 4.1 which will play a central role in this and other sections. For

smooth domains and smooth right hand sides (4.5) says that the usual "shift" theorem holds. If Ω is polygonal, then the regularity of u depends on the size of the maximal interior angle α_m. For simplicity, we have distinguished between two cases (convenient for our purposes), i.e., when Ω is convex and non-convex, respectively. In the first case (4.6) says that $u \in H^2(\Omega)$ for $f \in L_2(\Omega)$ and the "shift" theorem with $u \in H^2(\Omega)$ holds. Actually one can derive additional (and finer) results which depend on the angle α_m (in this case $\alpha_m < \pi$) and we refer the reader to Grisvard [1985a]. In the non-convex case it turns out that, in general, $u \notin H^2(\Omega)$ (even if $f \in C^\infty(\Omega)$). Roughly speaking the worst behavior of the solution is near the origin where $\pi < \alpha_m \leq 2\pi$ and where, for example, the solution of (4.2) admits the asymptotic expansion

$$u = a\rho^{\pi/\alpha_m} \sin(\pi/\alpha_m)\theta + \omega. \tag{4.6b}$$

Here ρ and θ are polar coordinates and $\omega \in H^2(\Omega)$. The coefficient a is called the stress intensity factor and is of importance to engineers. Obviously, $u \in H^2(\Omega^*)$ for any subdomain Ω^* of $\overline{\Omega}$ which excludes a neighborhood of the origin. It is easy to formally see that $\rho^{\pi/\alpha_m} \sin(\pi/\alpha_m)\theta$ does not in general have two derivatives in $L_2(\Omega)$ and that the weight ρ^t, for $t > 2(1-\pi/\alpha_m)$, is sufficient to insure convergence of the second derivatives in L_2. Analogous considerations hold for the Neumann Problem.

We shall need to know how well the interpolant u_I approximates a function u with finite weighted semi-norm $|u|_{2,\Omega,\rho,t}$. The following inequalities will be derived in Section 7 (Theorem 7.1).

<u>Lemma 4.1</u>. Let $\{\tau\}^h$ be a quasi-uniform triangulation of Ω. Then for any fixed $\epsilon > 0$

$$\|u-u_I\|_\Omega \le c_\epsilon h^{1+\pi/\alpha_m-\epsilon} |u|_{2,\Omega,t} \tag{4.7}$$

and

$$\|\nabla(u-u_I)\|_\Omega \le c_\epsilon h^{\pi/\alpha_m-\epsilon} |u|_{2,\Omega,t}, \tag{4.8}$$

where $t = 2(1-\pi/\alpha_m+\epsilon)$.

<u>4B. Galerkin Finite Element Discretizations of the Dirichlet and Neumann Problems, Error Estimates in $H^1(\Omega)$</u>.

The Galerkin finite element discretizations for Problems I and II are obtained by requiring that the corresponding weak formulations hold over the appropriate finite element space. Let us begin with Dirichlet's problem. From now on we shall assume that Ω is a polygonal domain, generalizations will be discussed at the end of this section.

Let $S_0^h(\Omega)$ be the space of continuous piecewise linear functions defined on quasi-uniform triangulation of Ω of size h. For each u, a solution of Dirichlet's Problem

(4.2), let $u_h \in S_0^h(\Omega)$ be the Galerkin approximation defined by

Problem IG: Find $u_h \in S_0^h(\Omega)$ satisfying

(4.9) $$(\nabla u_h, \nabla\varphi)_\Omega = (f, \varphi)_\Omega, \text{ for all } \varphi \in S_0^h(\Omega).$$

For each u, a solution of the Neumann Problem (4.4), let $u_h \in S^h(\Omega)$ be the Galerkin approximation defined as follows:

Problem IIG: Find $u_h \in S^h(\Omega)$ satisfying

(4.10) $$(\nabla u_h, \nabla\varphi)_\Omega + (u, \varphi)_\Omega = (f, \varphi)_\Omega, \text{ for all } \varphi \in S^h(\Omega).$$

Remark 4.2. Let $\varphi_i(x) \in S_0^h(\Omega)$ (resp. $S^h(\Omega)$), $i = 1, \ldots, N$, be the nodal basis functions discussed in Section 2. Then u_h may be written as

(4.11) $$u_n = \sum_{i=1}^{N} a_i \varphi_i,$$

and (4.9) (resp. (4.10)) leads to the linear system of algebraic equations

(4.12) $$\sum_{i=1}^{N} M_{ij} a_i \equiv \sum_{i=1}^{N} (\nabla\varphi_i, \nabla\varphi_j)_\Omega a_i = (f, \varphi_j)_\Omega, \; j = 1, \ldots, N.$$

Note that the matrix $M = \{M_{ij}\}$ is symmetric and sparse (as discussed in Section 3). This matrix is often referred to as the stiffness matrix.

Remark 4.3. First consider Problem IG. In view of (4.2) it follows that if u_h satisfies (4.9) then

$$(4.13) \qquad (\nabla(u-u_h),\nabla\varphi)_\Omega = 0 \quad \text{for all} \quad \varphi \in S_0^h(\Omega).$$

But (4.13) is just (3.13), i.e.,

$$(4.14) \qquad u_h = R_h u,$$

the best approximation to u in the $H_0^1(\Omega)$ norm discussed in Section 3. Hence existence, uniqueness and error estimates in $H^1(\Omega)$ follow immediately from Example 2 in Section 3. For Problem IIG, it follows from (4.4) and (4.10) that

$$(4.15) \qquad (\nabla(u-u_h),\nabla\varphi)_\Omega + (u-u_h,\varphi)_\Omega = 0 \quad \text{for all} \quad \varphi \in S^h(\Omega).$$

Thus, in this case,

$$(4.16) \qquad u_h = R_h^N u,$$

the best approximation to u in $S^h(\Omega)$ in the $H^1(\Omega)$ norm. Existence, uniqueness and estimates for Problem IIG follow from the discussion in Example 3 of Section 3. We shall collect these facts in the next Theorem and also treat the case where

$u \notin H^2(\Omega)$ (non-convex domains). Furthermore, for completeness, we shall derive the results concerning existence, uniqueness and basic energy estimates for the case of Problem IG without assuming the results of Section 3. We do this because one of the methods of proof goes over (with some modifications) to other problems where the approximation u_h is not the best approximation with respect to some norm.

<u>Theorem 4.2</u>. Let $\{\tau\}^h$ be a quasi-uniform triangulation of Ω, a polygonal domain, and $S_0^h(\Omega)$ be as in Section 2.

i) For each $f \in L_2(\Omega)$ (or $f \in (H_0^1(\Omega))'$), there exists a unique $u_h \in S_0^h(\Omega)$ satisfying (4.9).

ii) If u satisfies (4.2) and $u_h \in S_0^h(\Omega)$ satisfies (4.9), then $u-u_h$ satisfies (4.13). Thus $u = R_h u$, the best approximation in $S_0^h(\Omega)$ to u in the $H_0^1(\Omega)$ norm. Hence

$$\|\nabla(u-u_h)\|_\Omega = \inf_{\chi\in S_0^h(\Omega)} \|\nabla(u-\chi)\|_\Omega. \tag{4.17}$$

iii) If $f \in L_2(\Omega)$ and Ω is convex, then $u \in H_0^1(\Omega) \cap H^2(\Omega)$ and

$$\begin{aligned} \|\nabla(u-u_h)\|_\Omega &\le ch|u|_{2,\Omega} \\ &\le ch\|f\|_\Omega. \end{aligned} \tag{4.18}$$

iv) If $f \in L_2(\Omega)$ and Ω is non-convex as in part iii) of Theorem 4.1 then for any fixed $\epsilon > 0$

$$(4.19) \qquad \|\nabla(u-u_h)\|_\Omega \le c_\epsilon h^{\pi/\alpha_m - \epsilon} |u|_{2,\Omega,t}$$

$$\le c_\epsilon h^{\pi/\alpha_m - \epsilon} \|f\|_\Omega .$$

where $t = 2(1-\pi/\alpha_m+\epsilon)$.

We leave the reader with the exercise of deriving analogous results for Problem IIG.

Proof: i) There are several equivalent ways of proving existence and uniqueness for Problem IG. For example, $(\nabla u, \nabla\varphi)_\Omega$ is an inner product on $H_0^1(\Omega)$ hence on $S_0^h(\Omega) \subset H_0^1(\Omega)$. Since, for fixed f, $(f,\varphi)_\Omega$ is a bounded linear functional on $S_0^h(\Omega)$, the Riesz Representation Theorem implies that (4.9) has a unique solution for each f.

An alternate proof is as follows: By Remark 4.2, the equation (4.9) is equivalent to the linear system of algebraic equations (4.12). Existence and uniqueness of the problem will follow if we can prove that the homogeneous equation corresponding to (4.12) has only the trivial solution. This is equivalent to showing that the homogeneous equation $(f \equiv 0)$ corresponding to (4.9)

$$(\nabla u_h, \nabla\varphi)_\Omega = 0, \text{ for all } \varphi \in S_0^h(\Omega),$$

has only the trivial solution $u_h \equiv 0$. Choosing $\varphi = u_h$ we have $\|\nabla u_h\|_\Omega \equiv 0$, which, since $u_h = 0$ on $\partial\Omega$, implies that $u_h \equiv 0$.

ii) In order to derive the estimate (4.17), we note that

$$\|\nabla(u-u_h)\|_\Omega^2 = (\nabla(u-u_h), \nabla(u-u_h))_\Omega .$$

In view of the basic error equation (4.13), it follows that for any $\chi \in S_0^h(\Omega)$

$$\|\nabla(u-u_h)\|_\Omega^2 = (\nabla(u-u_h), \nabla(u-\chi))_\Omega$$

$$\leq \|\nabla(u-u_h)\|_\Omega \|\nabla u-\chi\|_\Omega ,$$

where we have used the Cauchy-Schwarz inequality. Hence

$$\|\nabla(u-u_h)\|_\Omega \leq \|\nabla(u-\chi)\|_\Omega ,$$

from which (4.17) easily follows. Hence $u_h = R_h u$ and the proof of ii) is complete.

iii) The inequality (4.18) follows from (2.4), (4.6) and (4.19) with the choice $\chi = u_I$.

iv) The inequality (4.19) follows from (4.8), (4.6a) and (4.17) with the choice $\chi = u_I$. This completes the proof of Theorem 4.2.

4C. Error Estimates in $L_2(\Omega)$, the Duality Argument of Aubin and Nitsche.

In view of the regularity of the solutions, the energy estimates given above for the finite element method are optimal with respect to the power of h. That is they are the best that we can expect in general from this subspace from approximation theory, even if the solution were explicitly known apriore. We now ask, how well does u_h approximate u in other norms? For example in $L^2(\Omega)$ (maximum norm estimates are discussed in Section 9). For a convex domain and $f \in L^2(\Omega)$, $u \in H^2(\Omega)$, and (2.4) says that for the interpolant, $\|u-u_I\|_\Omega \leq Ch^2|u|_{2,\Omega}$. We shall show, in this case, that this estimate also holds with the finite element solution u_h replacing u_I. As discussed above, if Ω is non-convex with $\pi < \alpha_m \leq 2\pi$, and $f \in L^2(\Omega)$, then (4.8) gives $\|u-u_I\|_\Omega = O(h^{1+\pi/\alpha_m-\epsilon})$. Unfortunately this estimate does not hold with u_I replaced by the finite element solution u_h. The best that can be proven (even if $f \in C^\infty(\Omega)$) is

$$\|u-u_h\|_\Omega = O(h^{2(\pi/\alpha_m-\epsilon)}).$$

Notice that since $\pi/\alpha_m < 1$, $\frac{2\pi}{\alpha_m} < 1 + \frac{\pi}{\alpha_m}$. Thus, in this case the Galerkin finite element method is not optimal in the $L_2(\Omega)$ norm in the sense of approximation theory. It has been proved that this estimate is sharp with respect to the power of h. Both of these estimates will be proved using a fundamental duality argument due to Aubin and Nitsche.

Theorem 4.3. Under the conditions of Theorem 4.2:

i) If Ω is convex and $f \in L^2(\Omega)$, then

(4.20) $$\|u-u_h\|_\Omega \lesssim Ch^2|u|_{2,\Omega} \lesssim Ch^2\|f\|_\Omega.$$

ii) If Ω is non-convex and $f \in L^2(\Omega)$ with α_m and t as in Theorem 4.2, then

(4.21) $$\|u-u_h\|_\Omega \lesssim C_\epsilon h^{2(\frac{\pi}{\alpha_m} - \epsilon)} |u|_{2,\Omega,\rho,t} \lesssim Ch^{2(\frac{\pi}{\alpha_m} - \epsilon)} \|f\|_\Omega.$$

<u>Proof</u>: Now

(4.22) $$\|u-u_h\|_\Omega^2 = (u-u_h, u-u_h)_\Omega.$$

We know that $u-u_h$ satisfies (4.13). The idea now is to connect the L^2 inner product with the $H^1(\Omega)$ inner product. This is essentially done by introducing an auxilliary function v which is the "weak" solution of $-\Delta v = u-u_h$ in Ω, $v = 0$ on $\partial\Omega$, and then integrating (4.8) by parts. Equivalently, let $v \in H_0^1(\Omega)$ satisfy

(4.23) $$(\nabla v, \nabla\varphi)_\Omega = (u-u_h, \varphi)_\Omega \quad \text{for all} \quad \varphi \in H_0^1(\Omega).$$

Now, $u-u_h \in L^2(\Omega)$ and hence, if Ω is convex, $v \in H^2(\Omega) \cap H_0^1(\Omega)$ and

(4.24) $$|v|_{2,\Omega} \lesssim C\|u-u_h\|_\Omega.$$

If Ω is non-convex, then we are only able to say that, with ρ and t as above,

(4.25) $$|v|_{2,\Omega,t} \leq C_t \|u-u_h\|_\Omega.$$

Using (4.23), with $\varphi = u-u_h$, and the basic orthogonality property (4.13) of $u-u_h$, it follows that for any $\chi \in S_0^h(\Omega)$

$$(u-u_h, u-u_h)_\Omega = (\nabla(u-u_h), \nabla v)_\Omega = (\nabla(u-u_h), \nabla(v-\chi))_\Omega.$$

Therefore

(4.26) $$\|u-u_h\|_\Omega^2 \leq \|\nabla(u-u_h)\|_\Omega \|\nabla(v-\chi)\|_\Omega.$$

Choosing $\chi = v_I$, it follows from (2.4), (4.24) and (4.26) that when Ω is convex

$$\|u-u_h\|_\Omega^2 \leq Ch\|\nabla(u-u_h)\|_\Omega |v|_{2,\Omega}$$

$$\leq Ch\|\nabla(u-u_h)\|_\Omega \|u-u_h\|_\Omega,$$

or

(4.27) $$\|u-u_h\|_\Omega \leq Ch\|\nabla(u-u_h)\|_\Omega.$$

In a similar fashion it follows from (4.8), (4.25) and (4.26) that when Ω is non-convex

(4.28) $$\|u-u_h\|_\Omega \leq C_\epsilon h^{\pi/\alpha_m-\epsilon} \|\nabla(u-u_h)\|_\Omega.$$

The inequalities (4.20) and (4.21) now follow from (4.27) and (4.28) respectively using Theorem 4.2.

Remark 4.4. Note that the increase in accuracy of the error in the L_2 norm as compared to that of the H^1 norm depends on the accuracy with which the solution of the auxilliary ("dual") problem (4.23) can be approximated in the H^1 norm. It will be of importance in Section 5 to notice that (4.27) and (4.28) may be derived if the solution $u_h \in S_0^h(\Omega)$ of Problem IIG is assumed to exist.

Remark 4.5. Theorems 4.2 and 4.3 can be easily generalized to more general second order differential operators. For simplicity consider the boundary value problem

$$(4.29)\qquad Lu = -\sum_{i,j=1}^{2} \frac{\partial}{\partial x_j}\left(a_{ij}\frac{\partial u}{\partial x_i}\right) + \sum_{j=1}^{2}\frac{\partial(b_j u)}{\partial x_j} + cu = f \text{ in } \Omega$$

$$u = 0 \text{ on } \partial\Omega.$$

Multiplying by φ and integrating by parts we obtain the weak formulation of (4.29);

Find $u \in H_0^1(\Omega)$ satisfying

$$(4.30)\qquad D(u,\varphi) = \sum_{i,j=1}^{2}\left(a_{ij}\frac{\partial u}{\partial x_i},\frac{\partial\varphi}{\partial x_j}\right)_\Omega - \sum_{j=1}^{2}\left(b_j u,\frac{\partial\varphi}{\partial x_j}\right)_\Omega + (cu,\varphi)_\Omega$$

$$= (f,\varphi)_\Omega \quad \text{for all } \varphi \in H_0^1(\Omega).$$

Let us assume that L is uniformly elliptic on Ω and that $D(\cdot,\cdot)$ is strictly coercive on $H_0^1(\Omega)$, i.e., there exists a constant $c > 0$ such that

$$C\|u\|_{1,\Omega}^2 \le D(u,u). \tag{4.31}$$

We associate with D the adjoint form D^* defined on $H_0^1(\Omega) \times H_0^1(\Omega)$

$$D^*(u,\varphi) = D(\varphi,u). \tag{4.32}$$

Then under the condition (4.31), the boundary value problem (4.17), and the adjoint problem: Find $v \in H_0^1(\Omega)$ satisfying for given $g \in L_2(\Omega)$

$$D^*(v,\varphi) = (g,\varphi) \quad \forall\varphi \in H_0^1(\Omega). \tag{4.19}$$

Both have unique solutions (by the Lax-Milgram Lemma). The regularity results i) and ii) of Theorem 4.1 still hold; however, (4.7) must be modified. Without going into details, a crude estimate which is valid is (4.7) with any $t > 1$.

The finite element method corresponding to (4.30) is: Find $u_h \in S_0^h(\Omega)$ satisfying

$$D(u_h,\varphi) = (f,\varphi) \quad \text{for all} \quad \varphi \in S_0^h(\Omega).$$

Existence follows from (4.31), and the Lax Milgram Lemma and estimates in $H^1(\Omega)$ as before. Estimates in $L_2(\Omega)$ follow the duality procedure as given in the proof of Theorem 4.3 except instead of solving the auxilliary problem (4.23) we solve the adjoint problem

$$D^*(v,\varphi) = (u-u_h,\varphi) \quad \text{for all} \quad \varphi \in H_0^1(\Omega).$$

Notice that in general $D(u,v)$ is not symmetric and hence not an inner product on $H_0^1(\Omega)$, so that u_h no longer has the interpretation as being a "best" approximation. However it is "good enough" in $H^1(\Omega)$.

5. Indefinite Second Order Elliptic Problems.

In the previous section, we studied boundary value problems whose weak formulation involved quadratic forms which were positive definite. Here we will consider problems where the corresponding quadratic forms are indefinite but satisfy a Gårding inequality. We shall show that the results for the finite element methods derived in the previous section still hold, provided the mesh size h is made sufficiently small. For simplicity we will restrict ourselves to a model problem, however the methods used here easily generalize to other problems and we refer the reader to Schatz [1974]. Some simple generalizations will be discussed at the end of this section.

5A. The Model Problem and Some Preliminaries.

Consider the model problem

$$(5.1)\qquad \begin{aligned} Lu &= -\Delta u - ku = f \quad \text{in } \Omega, \\ u &= 0 \quad \text{on } \partial\Omega. \end{aligned}$$

Here k is a constant and say $f \in L_2(\Omega)$. The weak formulation of (5.1) is

Problem III. Find $u \in H_0^1(\Omega)$ satisfying

$$(5.2)\qquad (\nabla u, \nabla\varphi)_\Omega - k(u,\varphi)_\Omega = (f,\varphi)_\Omega \quad \text{for all } \varphi \in H_0^1(\Omega).$$

The corresponding finite element discretization of (5.2) is

Problem IIIG. Find $u_h \in S_0^h(\Omega)$ satisfying

$$(5.3) \quad (\nabla u_h, \nabla\varphi)_\Omega - k(u_h,\varphi)_\Omega = (f,\varphi)_\Omega \quad \text{for all} \quad \varphi \in S_0^h(\Omega).$$

When $k \leq 0$, the results for this problem are precisely the same as for Problem IG, discussed in Section 4. Hence here we will only consider the case where $k > 0$. In order to insure that (5.2) has a unique solution, we shall assume that k is not an eigenvalue of $-\Delta$. Under this condition we have the following well known result.

Lemma 5.1. Suppose that for $f \equiv 0$, (5.2) has only the trivial solution $u \equiv 0$ in $H_0^1(\Omega)$. Then, for each $f \in L_2(\Omega)$, there exists a unique solution $u \in H_0^1(\Omega)$ of (5.2). Furthermore the regularity results of Theorem 4.1 hold.

5B. Existence and Error Estimates for the Finite Element Method.

Our problem here is to prove existence of a solution of (5.3) and then error estimates. We shall now prove

Theorem 5.1. Let Ω, $S_0^h(\Omega)$ and f be as in Theorems 4.1 and 4.2. Then if h is sufficiently small, there exists a unique solution $u_h \in S_0^h(\Omega)$ of (5.3). Furthermore the estimates (4.18), (4.19), (4.20) and (4.21) hold in this case.

Proof: Let $e = u-u_h \in H_0^1(\Omega)$, then

$$(5.4)\qquad \|\nabla e\|_\Omega^2 - k\|e\|_\Omega^2 = (\nabla e,\nabla e)_\Omega - k(e,e)_\Omega$$

$$= (\nabla e,\nabla(u-u_h))_\Omega - k(e,u-u_h)_\Omega .$$

Now let us suppose that $u \in H_0^1(\Omega)$ satisfies (5.1) and $u_h \in S_0^h(\Omega)$ satisfies (5.3), then e satisfies

$$(5.5)\qquad (\nabla e,\nabla\varphi)_\Omega - k(e,\varphi)_\Omega = 0 \quad \text{for all} \quad \varphi \in S_0^h(\Omega).$$

Hence from (5.4) and (5.5)

$$(5.6)\qquad \|\nabla e\|_\Omega^2 - k\|e\|_\Omega^2 = (\nabla e,\nabla u)_\Omega - k(e,u)_\Omega .$$

Using Schwarz's and the arithmetic geometric mean inequalities

$$\|\nabla e\|_\Omega^2 - k\|e\|_\Omega^2 \le \frac{1}{2}\|\nabla e\|_\Omega^2 + 2\|\nabla u\|_\Omega^2 + \frac{k}{2}\|e\|_\Omega^2 + 2k\|u\|_\Omega^2$$

or

$$(5.7)\qquad \|\nabla e\|_\Omega^2 - 3k\|e\|_\Omega^2 \le 4(\|\nabla u\|_\Omega^2 + k\|u\|_\Omega^2).$$

We next claim that the Aubin and Nitsche duality argument can be used to prove that under the assumption that u_h is such that $e = u-u_h$ satisfies (5.5) then we have the estimate (see Remark 4.4)

$$(5.8)\qquad \|e\|_\Omega \le ch^\gamma \|\nabla e\|_\Omega .$$

Here $\gamma = 1$ if Ω is convex, and $\gamma = \pi/\alpha_m - \epsilon$ if Ω is non-convex. Granting this inequality for a moment, let us prove Theorem 5.1. Using the estimate (5.8) is the key to the method for combining (5.7) and (5.8) we arrive at

$$(1-3kc^2h^{2\gamma})\|\nabla e\|_\Omega^2 \le 4(\|\nabla u\|_\Omega^2 + k\|u\|_\Omega^2).$$

Hence for h sufficiently small

$$0 < \beta = 1-3kc^2h^{2\gamma},$$

and

$$\|\nabla e\|_\Omega^2 \le \frac{4}{\beta}(\|\nabla u\|_\Omega^2 + k\|u\|_\Omega^2). \tag{5.9}$$

We are now in a position to prove existence of a solution of (5.3). Since (5.3) is equivalent to a finite system of linear algebraic equations for u_h (at the nodes), we need only show that the homogeneous equation has only the trivial solution. Let $f = 0$. This implies $u = 0$ and therefore from (5.9)

$$\|\nabla u_h\|_\Omega \le 0.$$

This in turn implies $u_h = 0$. Thus the solution of (5.3) exists and is unique provided h is sufficiently small. In order to prove the error estimates, we note that if $u \in S_0^h(\Omega)$ then $u_h = u$, i.e., the Galerkin method (5.3) is the identity on the subspace. Hence from (5.9)

$$\text{(5.10)} \qquad \|\nabla e\|_\Omega^2 \equiv \|\nabla(u-u_I)-\nabla(u_h-u_I)\|_\Omega^2 \le C(\|u-u_I\|_{1,\Omega}^2)$$

and the error estimates in $H^1(\Omega)$ follow immediately from the properties of the interpolant. The estimates in $L_2(\Omega)$ now follow immediately from (5.8) and (5.10). Thus the proof of Theorem 5.1 is complete once we have shown (5.8).

To prove (5.8) we employ the Aubin-Nitsche argument assuming (5.5) holds (cf. Remark 4.4). To start with

$$\|e\|_\Omega = \sup_{v \in L_2(\Omega)} \frac{(e,v)_\Omega}{\|v\|_\Omega}.$$

Let $\omega \in H_0^1(\Omega)$ satisfy the "adjoint" problem

$$(\nabla\omega,\nabla\varphi)_\Omega - k(\omega,\varphi)_\Omega = (v,\varphi)_\Omega \quad \text{for all} \quad \varphi \in H_0^1(\Omega).$$

From our assumption that this equation has a unique solution we have that Lemma 5.1 holds for this problem. Then using the regularity properties of ω and (5.5)

$$\begin{aligned}(e,v)_\Omega &= (\nabla e,\nabla\omega)_\Omega - k(e,\omega)_\Omega \\ &= (\nabla e,\nabla(\omega-\omega_I))_\Omega - k(e,\omega-\omega_I)_\Omega \\ &\le ch^\gamma(\|\nabla e\|_\Omega + \|e\|_\Omega)\|v\|_\Omega.\end{aligned}$$

Using Poincare's inequality (1.7)

$$(e,v) \le ch^\gamma(\|\nabla e\|_\Omega)\|v\|_\Omega.$$

and the desired result easily follows, which completes the proof.

5C. Some Generalizations.

Let us end this section by noting that Theorem 5.1 can be easily extended to the case where L is a general second order uniformly elliptic operator with sufficiently nice coefficients. In this case a Gårding inequality trivially holds, i.e.

$$c_1\|\nabla u\|_\Omega^2 - c_2\|u\|_\Omega^2 \leq (Lu,u)_\Omega \quad \text{for all} \quad u \in H_0^1(\Omega)$$

for some constants $c_1 > 0$ and c_2. The only basic change in the proof occurs in proving the duality estimate (5.8), which is handled as in Remark 4.5. Furthermore the technique in the proof can be used in many other problems in which a Gårding type inequality holds. We again refer the reader to Schatz [1974].

6. Local Error Estimates.

As discussed in Section 4, the finite element solution of (4.9) (resp. (4.10)) is the orthogonal projection of the solution of (4.2) (resp. (4.4)) into $S_0^h(\Omega)$. In general therefore it depends on the properties of u in all of Ω. Along these lines Theorem 4 gives estimates for the error which depends on the smoothness of u on all of Ω. In the case that Ω is non-convex and we are again using the piecewise linear subspaces $S_0^h(\Omega)$ (resp. $S^h(\Omega)$), we obtained for example for any $\epsilon > 0$

$$\|\nabla(u-u_h)\|_{L^2(\Omega)} \le C_\epsilon h^{(\pi/\alpha_m)-\epsilon} |u|_{2,\Omega,\rho,t},$$

where ρ and t and α_m are as in Theorem 4.2, $\pi < \alpha_m \le 2\pi$. Again note that $\pi/\alpha_m < 1$ and in general we cannot improve this estimate even if $f \in L^2(\Omega)$ or $C^\infty(\bar{\Omega})$. Again let us simplify matters a little by assuming that all the corners are convex except one, i.e., $0 < \alpha_1 \le \alpha_2 \le \dots \le \alpha_{m-1} < \pi$ and $\pi < \alpha_m \le 2\pi$. Now $u \in H^2(\tilde{\Omega})$ on any subdomain $\tilde{\Omega}$ which excludes a neighborhood of the corner with interior angle α_m. Here we are assuming that $f \in L^2(\Omega)$ (see Grisvard [1985a]). It is natural to ask whether the accuracy of the finite element solution is better in the region away from the corner where u is smooth? In this particular case the answer is yes; in fact the error will be of optimal order h except in the case when $\alpha_m = 2\pi$ when it is of order $h^{1-\epsilon}$, any $\epsilon > 0$. Another interesting situation occurs even when the boundary of Ω is smooth but f and hence u is not smooth in certain parts of Ω and smooth in others and again the same question may be asked. In this section we shall derive local error estimates for the finite element method, i.e., estimates on subdomains of Ω and see to what extent they are influenced by the solution u and the nature of the domain away from the subdomain. These estimates have been useful in a number of applications. They will also be useful in our derivation of maximum norm estimates given in a later section. For simplicity in presentation we shall give a slightly modified treatment of a special case of a result proved in Nitsche and Schatz [1974].

We start with some notation. For $x \in \bar{\Omega}$, $B(x,d)$ will denote the open ball of radius d centered at x and $B'(x,d) = B(x,d) \cap \Omega$. The basic error estimate is as follows:

<u>Theorem 6.1</u>. Let u and u_h satisfy (4.2) and (4.9), respectively. There exist positive constants C and C^* such that if h is sufficiently small and $d \geq C^*h$ then for any $\chi \in S_0^h$

$$(6.1)\quad \|\nabla(u-u_h)\|_{B'(x,d)}$$

$$\leq C\{\|\nabla(u-\chi)\|_{B'(x,2d)}+d^{-1}\|u-\chi\|_{B'(x,2d)}+d^{-1}\|u-u_h\|_{B'(x,2d)}\}.$$

Here C and C^* are independent of u, h, u_h, x, χ and d.

The proof of this result is lengthy. Let us first discuss its meaning, state a generalization, and then give an application to a specific problem. The estimate (6.1) may be interpreted as follows: The error in the norm $H^1(B'(x,d))$ can be bounded by three terms. The first two are just the best approximation error that the subspace S_0^h can provide on the larger subdomain $B'(x,2d)$ (which, for example, can be bounded by the local interpolation error in $H^1(B'(x,2d))$. Roughly speaking, the first term on the right is like the error in an H^1 projection of u just on $B'(x,2d)$. The third term is the error measured in the "weaker" norm on $L^2(B'(x,2d))$, which is multiplied by the inverse of the distance between the two domains. The effects on the error due to the nature of the boundary or solution outside of $B'(x,d)$ are still present in this latter term which must be estimated separately for each particular problem. One way to do this is to first use the inequality

$$\|u-u_h\|_{B'(x,2d)} \leq \|u-u_h\|_{\Omega},$$

and then estimate this latter term by the methods of the previous section. The significance of this weaker norm is that, under some important circumstances one can prove higher global rates of convergence in $L^2(\Omega)$ with relatively less requirements on the smoothness of u than one would need, for example in order to obtain the same rate of convergence in the $H^1(\Omega)$ norm.

Remark 6.1. Let us remark that the term $d^{-1}\|u-u_h\|_{B'(x,2d)}$ may be replaced by the weaker negative norm $d^{-p-1}\|u-u_h\|_{-p,B'(x,2d)}$

for any $p > 0$, (see Section 3C). Also, this result may be easily generalized to include more general subdomains of Ω. We shall do this now for two classes of subdomains the first of which will be useful in proving maximum norm estimates and the second so that we can give an easy application to the problem with a non-convex polynomial domain mentioned above.

Consider the concentric annuli

$$A_1 = \{x : \frac{d}{2} < |x-x_0| < \frac{3d}{2}\},\ A_2 = \{x : \frac{d}{4} \le |x-x_0| \le 2d\}$$

and $A_1' = A_1 \cap \Omega$, $A_2' = A_2 \cap \Omega$.

<u>Corollary 6.1</u>. The results of Theorem 6.1 hold with A_1' and A_2' replacing $B'(x,d)$ and $B'(x,2d)$ respectively.

<u>Proof</u>: Cover A_1 with a fixed number (independent of d) of balls $B(x,\frac{d}{8})$ of radius say $d/8$. Then also consider the balls $B(x_i,\frac{d}{4})$. Apply Theorem 6.1 to each pair $B'(x_i,\frac{d}{8})$ and $B'(x_i,\frac{d}{4})$. The desired result is easily obtained by appropriately summing.

Now let $\Omega_1 \subset \Omega_2$ be any fixed subdomains of Ω. Let $\Gamma_1 = \partial\Omega_1 \backslash (\partial\Omega_1 \cap \partial\Omega)$, $\Gamma_2 = \partial\Omega_2 \backslash (\partial\Omega_2 \cap \partial\Omega)$, i.e., Γ_1 and Γ_2 are the parts of $\partial\Omega_1$ and $\partial\Omega_2$ respectively which are not common to $\partial\Omega$. Suppose the $\mathrm{dist}(\Gamma_1,\Gamma_2) \ge d_0$ for some fixed d_0. Then

<u>Corollary 6.2</u>. With Ω_1 and Ω_2 as above Theorem 6.1 holds with Ω_1 and Ω_2 replacing $B'(x,d)$ and $B'(x,2d)$ respectively. C in general depends on Ω_0 and Ω_1.

We leave the proof to the reader. Before proving Theorem 6.1 let us apply Corollary 6.2 to the following concrete problem.

<u>Example</u>. Let Ω be the slit domain given in Figure 6.1 and let u and u_h satisfy (4.2) and (4.9) respectively.

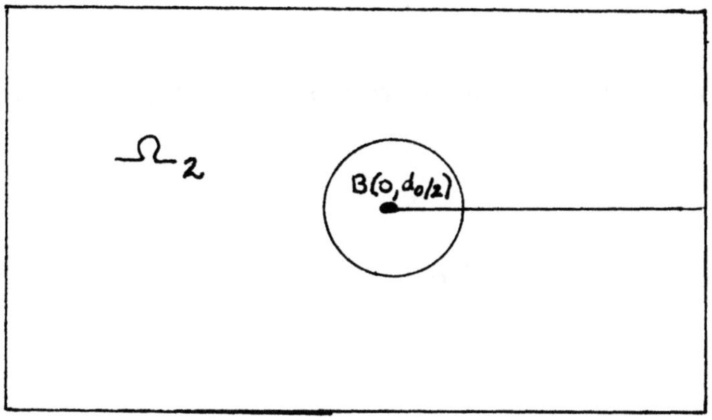

Figure 6.1

In this case the results of Theorems 4.2 and 4.3 yield the global estimates.

$$\|\nabla(u-u_h)\|_{L^2(\Omega)} \le C_\epsilon h^{1/2-\epsilon} |u|_{2,\Omega,t} \tag{6.2}$$

$$\|(u-u_h)\|_{L^2(\Omega)} \le C_\epsilon h^{1-2\epsilon} |u|_{2,\Omega,t} \tag{6.3}$$

for any $\epsilon > 0$.

Let us apply Corollary 6.2 where we take $\bar{\Omega}_1 = \bar{\Omega}\backslash B(0,d_0)$, $\bar{\Omega}_2 = \bar{\Omega}\backslash B(0,do/2)$ for any fixed $d_0 > 0$. Now $u \in H^2(\Omega_3)$, where $\bar{\Omega}_3 = \bar{\Omega}/B(0,do/4)$.

For h sufficiently small there exist a mesh domain D_h with $\Omega_2 \subset \bar{D}_h \subset \Omega_3$. Let $\chi = u_I$ and apply (2.4) to obtain

$$\|\nabla(u-\chi)\|_{\Omega_2} + d_0^{-1}\|u-\chi\|_{\Omega_2} \leq ch|u|_{2,\Omega_3}.$$

Combining this estimate with (5.3) and Corollary 5.2 yields the result

$$(6.4) \qquad \|\nabla(u-u_h)\|_{L^2(\Omega_1)} \leq C_\epsilon h^{1-2\epsilon}(|u|_{2,\Omega_3} + |u|_{2,\Omega,\rho,t})$$

for any $\epsilon > 0$. Comparing this with (6.2), this shows that convergence for the gradients in $L^2(\Omega_1)$ in the finite element method is better away from the corners than on the whole domain.

Proof of Theorem (6.1). The proof can be divided into two steps. First we will locally project u into the subspace. The first two terms on the right of (6.1) are essentially a bound for the error in this local projection. We then obtain an estimate for the difference between this local projection and u_h, which has certain nice properties.

For technical reasons we shall need 5 nested domains Ω_j defined by $\Omega_j = B'(x,(1+j/4)d)$, $j = 0,\ldots,4$. Note that $\Omega_0 = B'(x,d)$ and $\Omega_4 = B'(x,2d)$. We localize u by employing a smooth "cut-off" function w satisfying $w \in C_0^\infty(\Omega_3)$, $w \geq 0$,

$w \equiv 1$ on Ω_2 and $|D^\alpha w| \le Cd^{-|\alpha|}$. Such a function is easily constructed. Consider the function wu which vanishes outside of Ω_3 and let $P(wu) \in S_0^h(\Omega_4)$ be the projection of wu in $H^1(\Omega_4)$ defined by

(6.5) $$(\nabla wu-\nabla P(wu),\nabla\varphi) = 0 \quad \forall\ \varphi \in S_0^h(\Omega_4).$$

Since $w \equiv 1$ on Ω_2,

(6.6) $$\|\nabla u-\nabla u_h\|_{\Omega_0} \equiv \|\nabla(wu)-\nabla u_h\|_{\Omega_0}$$

$$\le \|\nabla(wu)-\nabla P(wu)\|_{\Omega_0}+\|\nabla P(wu)-\nabla u_h\|_{\Omega_0}$$

$$= I_1+I_2.$$

To estimate I_1 we use (3.14) with Ω_4 replacing Ω, the choice $\chi = 0$, and the properties of w to obtain

(6.7) $$I_1 \le \|\nabla(wu)-\nabla P(wu)\|_{\Omega_4} \le C\|\nabla(wu)\|_{\Omega_3}$$

$$\le C(\|\nabla u\|_{\Omega_3}+d^{-1}\|u\|_{\Omega_3}).$$

Set $V_h = P(wu)-u_h$. In order to estimate I_2 first notice that since u_h satisfies

(6.8) $$(\nabla u-\nabla u_h,\nabla\varphi)_\Omega = 0 \quad \forall\ \varphi \in S_0^h(\Omega)$$

and $wu = u$ on Ω_2, then subtracting (6.8) from (6.5) yields

$$(\nabla V_h, \nabla\varphi) = 0 \quad \forall\ \varphi \in S_0^h(\Omega_2). \tag{6.8a}$$

Such a $V_h \in S_0^h(\Omega_2)$ is called a "discrete harmonic function" on Ω_2. As we shall see it has certain properties similar to those of a harmonic function. In particular we shall show in our situation that the following holds:

<u>Lemma 6.1</u>. Let V_h be discrete harmonic on Ω_2, i.e., satisfy (6.8a), then there exists a constant C such that

$$\|\nabla V_h\|_{\Omega_0} \le Cd^{-1}\|V_h\|_{\Omega_2}. \tag{6.9}$$

Here C is independent of h, V_h, χ and d.

Assuming for a moment that we have proved this result, let us complete the proof of (6.1). Using (6.9) and the triangle inequality

$$\begin{aligned} I_2 &\le Cd^{-1}\|P(wu)-u_h\|_{\Omega_2} \\ &\le Cd^{-1}(\|P(wu)\|_{\Omega_2}+\|u-u_h\|_{\Omega_2}+\|u\|_{\Omega_2}). \end{aligned} \tag{6.10}$$

Since $P(wu) \in S_0^h(\Omega_4) \subset H^1(\Omega_4)$, where $\mathrm{diam}(\Omega_4) \le 4d$, then using Poincaré's inequality (1.7) and the properties of $P(wu)$. Next, we see that

$$(6.11)\qquad d^{-1}\|P(\omega u)\|_{\Omega_4} \le C\|\nabla P(\omega u)\|_{\Omega_4} \le C\|\nabla(\omega u)\|_{\Omega_3}$$

$$\le C(\|\nabla u\|_{\Omega_3}+d^{-1}\|u\|_{\Omega_3}).$$

Combining (6.11), (6.10), (6.7) and (6.6) we arrive at

$$(6.12)\qquad \|\nabla(u-u_h)\|_{\Omega_0} \le C(\|\nabla u\|_{\Omega_4}+d^{-1}\|u\|_{\Omega_4}+d^{-1}\|u-u_h\|_{\Omega_4}).$$

Now notice that for any $\chi \in S_0^h(\Omega)$

$$(\nabla u-\nabla u_h,\nabla\varphi) = (\nabla(u-\chi)-\nabla(u_h-\chi),\nabla\varphi) = 0,$$

i.e., $u_h-\chi$ is the projection of $u-\chi$. Applying the estimate (6.12) to this pair of function

$$\|\nabla(u-u_h)\|_{\Omega_0} \le C(\|\nabla(u-\chi)\|_{\Omega_4}+d^{-1}\|u-\chi\|_{\Omega_4}+d^{-1}\|u-u_h\|_{\Omega_4})$$

which is (6.1). Thus only the proof of (6.9) remains.

<u>Proof of Lemma 6.1</u>. Let ψ be a smooth "cut-off" function satisfying $\psi \in C_0^\infty(\Omega_1)$, $\psi \ge 0$, $\psi \equiv 1$ on Ω_0 and $|D^\alpha\psi| \le Cd^{-|\alpha|}$ for $|\alpha| \le 2$. Then

$$(6.13)\qquad \|\nabla V_h\|^2_{\Omega_0} \le \|\psi\nabla V_h\|^2_{\Omega_1} = \int_{\Omega_1} \psi^2\nabla V_h\cdot\nabla V_h dx$$

$$= \int_{\Omega_1} \nabla V_h\cdot\nabla(\psi^2 V_h)dx - \int_{\Omega_1} \nabla V_h\cdot 2\psi V_h\nabla\psi dx$$

$$= J_1+J_2.$$

Since ψ vanishes outside of Ω_1,

$$(6.14)\qquad |J_2| = |\int_\Omega \psi\nabla V_h\cdot 2V_h\nabla\psi dx| \le \|\psi\nabla V_h\|\cdot Cd^{-1}\|V_h\|_{\Omega_1}$$

$$\le \tfrac{1}{4}\|\psi\nabla V_h\|^2_{\Omega_1}+Cd^{-2}\|V_h\|^2_{\Omega_1},$$

where we have first used the Cauchy-Schwarz inequality and then the arithmetic-geometric mean inequality.

In order to estimate J_1 we note that, by taking C^* and hence d sufficiently large, there exists a mesh domain D_h (i.e., $\bar{D}_h = \cup\bar{T}_i$, $T_i \in \pi_h$) such that $\Omega_1 \subset D_h \subset \Omega_2$. The interpolant $(\psi^2 V_h)_I \in S^h_0(D_h) \subset S^h_0(\Omega_2)$ and since V_h satisfies (6.8a),

$$|J_1| = |\int_\Omega \nabla V_h\cdot\nabla(\psi^2 V_h-(\psi^2 V_h)_I)dx|$$

$$\le \|\nabla V_h\|_{D_h}\|\nabla(\psi^2 V_h-(\psi^2 V_h)_I\|_{D_h}.$$

The key to proving estimate (6.9) is the "super-approximation" property. Applying this property to the second term on the right it follows that

$$(6.15)\qquad |J_1| \le C\|\nabla V_h\|_{D_h} h(d^{-1}\|\psi\nabla V_h\|_{D_h} + d^{-2}\|V_h\|_{D_h}).$$

Using the arithmetic-geometric mean inequality and then the inverse property

$$\|\psi\nabla V_h\|_{D_n} Chd^{-1}\|\nabla V_h\|_{D_n} \le \tfrac{1}{4}\|\psi\nabla V_h\|^2_{D_n} + Ch^2d^{-2}\|\nabla V_h\|^2_{D_n}$$

$$\le \tfrac{1}{4}\|\psi\nabla V_h\|^2_{D_n} + Cd^{-2}\|V_h\|^2_{D_n}.$$

Again by the inverse property

$$Chd^{-2}\|\nabla V_h\|_{D_n}\|V_h\|_{D_n} \le Cd^{-2}\|V_h\|^2_{D_n}.$$

Using these last two inequalities in (6.15) and combining the result with (6.14) and (6.13) we obtain

$$\|\psi\nabla V_h\|^2_{D_n} \le \tfrac{1}{2}\|\psi\nabla V_h\|^2_{D_n} + cd^{-2}\|V_h\|^2_{D_n}.$$

Therefore

$$\|\nabla V_h\|^2_{\Omega_0} \le \|\psi\nabla V_h\|^2_{D_n} \le cd^{-2}\|V_h\|^2_{D_n} \le cd^{-2}\|V_h\|_{\Omega_2},$$

which completes the proof.

7. An Introduction to Mesh Refinement.

As seen in Section 4, the global errors in $H^1(\Omega)$ and $L^2(\Omega)$, for Galerkin's method for Dirichlet's problem on a domain Ω with a non-convex corner having maximal interior angle $\pi < \alpha_m \leq 2\pi$, are of order $h^{\pi/\alpha_m-\epsilon}$ and $h^{2(\pi/\alpha_m-\epsilon)}$, respectively, if a quasi-uniform mesh of size h is used. Again note that $\pi/\alpha_m < 1$ and $2\pi/\alpha_m < 2$. This is just one example where non-optimal rates of convergence are encountered due to the fact that the solution is not smooth.

A systematic study of mesh refinement for improving the accuracy of the approximation of solutions of second order elliptic boundary value problems is well beyond the scope of these lectures. This is because of the large variety of singular behavior that could occur. Here, we shall essentially restrict our attention to the above model problem to show how to construct an economical mesh so as to produce optimal global errors in $H^1(\Omega)$ and $L^2(\Omega)$ (and $C(\bar{\Omega})$), i.e., the same as if the solution were smooth. The methods discussed here can be used to study other problems with different "singular" behavior. Self-adaptive methods will not be discussed, rather, a mesh well designed on the basis of some a priori knowledge of the behavior of the solution. It is important to note that the precise form of the solution will not be needed (see the assumptions in Section 7A) even though this is known for our model problem. As previously stated, in this case u is of the form

$$u = a\left[\rho^{\pi/\alpha_m} \sin(\pi/\alpha_m)\theta\right]+\omega, \tag{7.1}$$

where $\rho = |x|$ and for $f \in L_2(\Omega)$, $\omega \in H^2(\Omega)$. It is the first term on the right of (7.1) that causes the difficulty (i.e., is not smooth). Throughout it will be assumed that $a \neq 0$ and, for further simplicity, that $\omega \in C^2(\bar{\Omega})$. Our first task will be to derive error estimates in various norms for the interpolants of a class of functions containing those functions of the form (7.1).

Then refined meshes will be constructed so that the interpolant achieves optimal order accuracy in $L_2(\Omega)$, $H^1(\Omega)$ and $C(\Omega)$. It is also important to note that the most economical meshes are different in each case. These results will then be applied to study the behavior of the error for our model Dirichlet problem. Even in this case the results in general are not simple. For estimates in $L_2(\Omega)$ and $H^1(\Omega)$ with refined grids we refer the reader to Babuska and Rosenzweig [1973] and for estimates in maximum norm to Schatz and Wahlbin [1979].

7A. Approximation Theory Revisited.

We begin by deriving some error estimates for the interpolant under somewhat more general smoothness conditions on the approximated function than those in Section 2.

Throughout this section it will be assumed that a triangulation $\{\tau\}$ of Ω is given consisting of triangles τ satisfying: To each $\tau \in \{\tau\}$ there exists an $h_\tau > 0$ such that τ is contained in a ball of radius h_τ and contains a ball of radius ch_τ, where $0 < c < 1$ is independent of τ.

<u>Lemma 7.1</u>. Let $\tau \in \{\tau\}$ and $u \in W^{2,1}(\tau)$, then

(7.2) $$\|u-u_I\|_{0,\tau} \leq ch_\tau |u|_{W^{2,1}(\tau)}.$$

(7.3) $$\|\nabla(u-u_I)\|_{0,\tau} \leq C|u|_{W^{2,1}(\tau)}$$

(7.4) $$\|u-u_I\|_{C(\bar{\tau})} \leq C|u|_{W^{2,1}(\tau)}.$$

Since the proof procedes along the lines of the proof of Theorem 2.1, we shall only outline the steps. First we map τ to a standard unit size triangle τ^* via an affine mapping. Denoting by u^* and u_I^* the transformed function u and its interpolant u_I, respectively, we have from the Sobolev inequalities (1.10) and (1.9)

$$\|u^*-u_I^*\|_{1,\tau^*} \leq c\|u^*-u_I\|_{W^{2,1}(\tau^*)}$$

and

$$\|u^*-u_I^*\|_{C(\bar{\tau}^*)} \leq c\|u^*-u_I^*\|_{W^{2,1}(\tau^*)}.$$

Since $u^*-u_I^*$ vanishes for all linear functions u^*, it follows that

$$\|u^*-u_I^*\|_{W^{2,1}(\tau^*)} \leq C|u^*|_{W^{2,1}(\tau^*)}.$$

Using this last inequality in the previous three inequalities and transforming back to τ, the inequalities (7.2), (7.3) and (7.4) easily follow.

<u>Remark 7.1</u>. Theorem 2.1 can also be obtained from Lemma 7.1. In fact if $u \in H^2(\tau)$ and $h_\tau = h$ then the Schwarz inequality yields

$$|u|_{W^{2,1}(\tau)} \leq ch|u|_{2,\tau}.$$

Hence (7.2) and (7.3) become

$$\|u-u_I\|_{0,\tau} \leq ch^2|u|_{2,\tau}$$

and

$$\|\nabla(u-u_I)\|_{0,\tau} \leq ch|u|_{2,\tau}.$$

Taken together these inequalities imply (2.2). If $u \in C^2(\bar{\tau})$, then

$$|u|_{W^{2,1}(\tau)} \leq ch^2|u|_{C^2(\bar{\tau})}$$

and (2.3) easily follows from (7.4).

We will now consider the approximation of a class of functions which have properties in common with those of (7.1). In what follows, it will be assumed for simplicity that $u \in C(\bar{\Omega}) \cap C^2(\bar{\Omega}-\{0\})$ and for some $0 < \gamma < 1$ satisfies

$$(7.5)\qquad |D^\alpha u| \le c\rho^{\gamma-2} \quad \text{for} \quad |\alpha| = 2, \text{ for all } \ x \in \Omega.$$

Note trivially that

$$(7.6)\qquad \sup_{x\in\Omega} |\rho^{2-\gamma} D^\alpha u(x)| \le M, \ |\alpha| = 2$$

and

$$(7.7)\quad \Big[\sum_{|\alpha|=2} \int_\Omega \rho^t (D^\alpha u)^2 dx \Big]^{1/2} = |u|_{2,\Omega,t} < M_t \ \text{ for } \ t > 2(1-\gamma).$$

Remark 7.2. If we take $\gamma = \pi/\alpha_m$ in (7.5), then functions of the form (7.1) satisfy (7.5). Other types of boundary value problems, for example some with equations with discontinuous coefficients, have solutions satisfying (7.5) where γ may be very close to zero.

Lemma 7.2. Let $\tau \in \{\tau\}$, $d_\tau = \mathrm{dist}(\bar{\tau}, 0)$ and let u satisfy (7.5), then for any $\epsilon > 0$

$$(7.8)\qquad \|u-u_I\|_\tau \le c_\epsilon \min(h_\tau^{1+\gamma-\epsilon}, h_\tau^2 d_\tau^{-t/2}) |u|_{2,\tau,t},$$

$$(7.9)\qquad \|\nabla(u-u_I)\|_\tau \le c_\epsilon \min(h_\tau^{\gamma-\epsilon}, h_\tau d_\tau^{-t/2}) |u|_{2,\tau,t}$$

for $t = 2(1-\gamma+\epsilon)$ and

$$(7.10)\qquad \|u-u_I\|_{C(\bar{\tau})} \le c\,\min(h_\tau^\gamma, h_\tau^2 d_\tau^{\gamma-2}) \sum_{|\alpha|=2} \sup_{x\in\tau} |\rho^{2-\gamma} D^\alpha u|.$$

<u>Proof</u>: We begin by estimating the $W^{2,1}(\tau)$ semi-norm of u. If $d_\tau > 0$, then for any α with $|\alpha| = 2$

$$(7.11)\qquad \int_\tau |D^\alpha u|dx \le \left[\int_\tau \rho^{-t}dx\right]^{1/2}\left[\int_\tau \rho^t |D^\alpha u|^2 dx\right]^{1/2}$$

$$\le c h_\tau d_\tau^{-t/2}\left[\int_\tau \rho^t |D^\alpha u|^2 dx\right]^{1/2}.$$

Similarly if $d_\tau = 0$, then

$$(7.12)\qquad \int_\tau |D^\alpha u|dx \le c h^{\gamma-\epsilon}\left[\int_\tau \rho^t |D^\alpha u|^2 dx\right]^{1/2}.$$

The inequalities (7.8) and (7.9) follow from (7.2), (7.3), (7.11) and (7.12). In order to prove (7.10) we use (7.4) and the fact that if $d_\tau > 0$

$$\int_\tau |D^\alpha u|dx \le c h_\tau^2 d_\tau^{\gamma-2} \sup_{x\in\tau} |\rho^{2-\gamma} D^\alpha u|.$$

Furthermore, if $d_\tau = 0$

$$\int_\tau |D^\alpha u|d\tau \le c h^\gamma \sup_{x\in\tau} |\rho^{2-\gamma} D^\alpha u|,$$

which completes the proof.

Lemma 7.2 is the basis for our next two approximation results. The first one deals with the case in which the mesh is quasi-uniform of size h and contains the results (4.7) and (4.8) which were used in estimating the error for the Dirichlet problem with a non-convex corner.

<u>Theorem 7.1</u>. Let u satisfy (7.5) with $0 < \gamma < 1$ and $h_\tau = h$, i.e., the mesh is quasi-uniform of size h, then for each $\epsilon > 0$ and $t = 1-\gamma+\epsilon$

$$\|u-u_I\|_\Omega \le ch^{1+\gamma-\epsilon}|u|_{2,\Omega,t}, \tag{7.13}$$

$$\|\nabla(u-u_I)\|_\Omega \le ch^{\gamma-\epsilon}|u|_{2,\Omega,t} \tag{7.14}$$

and

$$\|u-u_I\|_{C(\pi)} \le ch^\gamma \sum_{|\alpha|=2} \sup_{x\in\Omega}|\rho^{2-\gamma}D^\alpha u|. \tag{7.15}$$

<u>Proof</u>: In order to prove (7.13) we divide the mesh into two parts. Let Ω_0 be the union of the set of triangles which have a vertex at the origin and $\Omega_1 = \Omega/\overline{\Omega}_0$. Note that $\mathrm{dist}(\overline{\Omega}_1, 0) > ch$.

If $\tau \subset \Omega_0$, then (7.8) becomes

$$\|u-u_I\|_\tau \le ch^{1+\gamma-\epsilon}|u|_{2,\tau,\rho,t}. \tag{7.16}$$

If $\tau \subset \Omega_1$, then $d_\tau \ge ch$ and it again follows from (7.8) that (7.16) holds. The inequality (7.13) now easily follows by squaring both sides of (7.16) and summing over the τ. The proofs of (7.14) and (7.15) follow in the same manner and is left to the reader.

7B. Mesh Refinements for Optimal Order Accuracy in Various Norms for the Interpolant.

It is easy, based on Lemma 7.1, to create meshes so that optimal order accuracy can be achieved for the interpolants of functions satisfying (7.5) in $L^2(\Omega)$, $H^1(\Omega)$ and $C(\bar{\Omega})$. In Section 7C we will apply the results of the following theorem to a model Dirichlet problem.

Theorem 7.2. Let $h > 0$, $\epsilon > 0$, $0 < \gamma < 1$ and u satisfying (7.5). Then the following hold:

i) (An optimal mesh in $L_2(\Omega)$.) Let $\tau \in \{\tau\}$ and

$$h_\tau \le \begin{cases} ch^{\frac{2}{1+\gamma-\epsilon}} & \text{if } d_\tau = 0 \\ ch\, d_\tau^{\frac{1-\gamma+\epsilon}{2}} & \text{if } d_\tau > 0 \end{cases} . \tag{7.17}$$

Then

$$\|u-u_I\|_\Omega \le ch^2 |u|_{2,\Omega,t}. \tag{7.18}$$

ii) (An optimal H_1 mesh.) Let $\tau \in \{\tau\}$ and

$$(7.19) \qquad h_\tau \le \begin{cases} ch^{\frac{1}{\gamma-\epsilon}} & \text{if} \quad d_\tau = 0 \\ ch\, d_\tau^{1-\gamma+\epsilon} & \text{if} \quad d_\tau = 0 \end{cases} .$$

Then

$$(7.20) \qquad \|\nabla(u-u_I)\|_\Omega \le ch|u|_{2,\Omega,t} .$$

iii) (A optimal mesh in $C(\bar{\Omega})$.) If $\tau \in \{\tau\}$ and

$$(7.21) \qquad h_\tau \le \begin{cases} ch^{\frac{2}{\gamma}} & \text{if} \quad d_\tau = 0 \\ ch\, d_\tau^{1-\gamma/2} & \text{if} \quad d_\tau > 0 \end{cases} ,$$

then

$$(7.22) \qquad \|u-u_I\|_{C(\bar{\Omega})} \le ch^2 \sum_{|\alpha|=2} \sup_{x\in\Omega} |\rho^{\gamma-2} D^\alpha u| .$$

Let us postpone the proof of Theorem 7.2 for a moment and make several important remarks.

<u>Remark 7.3</u>. There are many different ways to construct meshes which satisfy (7.17), (7.19) and (7.21). Let us illustrate a naive way in the case of (7.21). Let $d_0 = ch^{2/\gamma}$. Draw a circle of radius d_0 about the origin and triangulate the region $G_0 = \{x:\rho \le d_0\} \cap \Omega$ with triangles of size $ch^{2/\gamma}$.

Then for each $j = 1,2,3$, etc., set $d_j = d_{j-1}+ch(d_{j-1})^{1-\gamma/2}$ and triangulate the region $G_j = \{x: d_{j-1} \leq \rho \leq d_j\} \cap \Omega$ with triangles of size $h(d_{j-1})^{1-\gamma/2}$. Do this until d_j reaches a maximum of 1 and then triangulate the rest of the region with triangles of size h.

<u>Remark 7.4</u>. One can show that the meshes described by (7.17), (7.19) and (7.21), respectively, are in some sense minimal in order to achieve optimal accuracy in their respective norms. Notice that they are very different. This is easily seen even from the size of the elements near the origin. In fact, a more refined grid necessary to give an optimal interpolation error in $C(\bar{\Omega})$ than $H^1(\Omega)$ and the $H^1(\Omega)$ refinement is necessarily finer than the optimal $L^2(\Omega)$ refinement. Obviously (7.21) => (7.19) => (7.17). Thus the optimal maximum norm mesh will give optimal interpolation errors in $H^1(\Omega)$ and $L_2(\Omega)$, and an optimal H^1 mesh will give an optimal interpolation error in $L_2(\Omega)$ but not vice-versa.

<u>Remark 7.5</u>. It is an easy exercise to show that the total number of triangles in all of the above triangulations is of the order ch^{-2}, i.e., proportional to the number of nodes in a quasi-uniform mesh.

<u>Proof of Theorem 7.2</u>. The proofs of (7.18), (7.20) and (7.22) follow in essentially the same manner from (7.8), (7.9) and (7.10), respectively. Here we will prove (7.20) and leave the others as an exercise for the reader.

If $\tau \in \{\tau\}$ and $d_\tau = 0$, then using (7.19) in (7.9) with $h_\tau = h^{\frac{2}{1-\gamma+\epsilon}}$, we obtain

$$\|\nabla(u-u_I)\|_\tau \leq ch^2|u|_{2,\tau,t}. \tag{7.23}$$

On the other hand if $d_\tau > 0$, then using $h_\tau = hd_\tau^{1-\sigma/2+\epsilon}$ in (7.9) it follows that (7.23) still holds. The inequality (7.20) now follows by squaring both sides of (7.23) and summing over all triangles.

7C. Refined Grids for Dirichlet's Problem with a Non-Convex Corner.

We now consider analyzing the error when grid refinement is used for Problem IG in order to obtain optimal order accuracy. Since u_h is the best approximation in $S_0^h(\Omega)$ to u in the H^1 norm, we shall start by placing a mesh which allows us to obtain optimal order accuracy for the interpolant. It turns out to be easy to show that one obtains $O(h)$ accuracy in H^1 and $O(h^2)$ in $L_2(\Omega)$ by using duality. It also turns out that the mesh defined by (7.17) which yields optimal $O(h^2)$ in L_2 for the interpolant does not yield optimal accuracy for the Galerkin solution (cf. Remark 7.6).

Theorem 7.3. Let Ω be non-convex as in Theorem 4.1 and $\{\tau\}$ be a triangulation of Ω satisfying (7.19), then

$$(7.24)\qquad \|\nabla(u-u_h)\|_\Omega \le ch^2|u|_{2,\Omega,\rho,t} \le ch\|f\|_\Omega$$

and

$$(7.25)\qquad \|u-u_h\|_\Omega \le ch^2|u|_{2,\Omega,t} \le ch\|f\|_\Omega.$$

Proof: The inequality (7.24) follows immediately from the best approximation property which implies

$$\|\nabla(u-u_h)\|_\Omega \le \|\nabla(u-u_I)\|_\Omega$$

and (7.20).

Following the proof of Theorem 4.3, the inequality (7.25) follows easily in view of the inequality

$$\|\nabla(v-v_I)\|_\Omega \le ch\|u-u_h\|_\Omega.$$

Remark 7.6. We invite the reader to show that if a grid is used satisfying (7.17) which is optimal in $L_2(\Omega)$ for the interpolant then

$$\|u-u_h\|_\Omega \le ch^{\frac{4\pi/\alpha_m}{1+\pi/\alpha_m}}\|f\|_\Omega.$$

where $\frac{4\pi/\alpha_m}{1+\pi/\alpha_m} < 2$. Hence non-optimal rates are obtained for the finite element method in $L_2(\Omega)$. In order to prove this estimate is sharp note that

$$\|\nabla(u-u_h)\|_\Omega^2 = (\nabla(u-u_h),\nabla u-u_h)_\Omega = (\nabla(u-u_h),\nabla u)_\Omega$$

$$= (u-u_h,\nabla u)_\Omega$$

or

$$\|\nabla(u-u_h)\|_\Omega^2 \le \|u-u_h\|_\Omega \|f\|_\Omega .$$

Hence for $f \neq 0$ $\|u-u_h\|_\Omega$ is at least the square of the error in $H^1(\Omega)$. We leave the remainder for the reader.

8. Maximum Norm Estimates for the L_2 Projection; A Method Using Weighted Norms.

In this section we shall derive maximum norm estimates for the L_2 projection onto $S^h(\Omega)$. There are basically several methods which have been developed for solving this problem. Here we shall prove our results via one of them, namely using certain weighted L_2 norms. The main ingredient of the proof is the super-approximation property (Section 2E). Maximum norm estimates for the much harder case of the Ritz projection will be considered in Section 9. We hope that this section will serve as an introduction to the weighted norms used in the proof of maximum norm estimates. As before throughout this section Ω will be a polygonal domain and $S^h(\Omega)$ will be the continuous piecewise linear functions defined on a quasi-uniform triangulation of Ω.

For $u \in C(\bar{\Omega})$ let $u_h = P_h u \in S^h(\Omega)$ be the L_2 projection satisfying

$$(8.1) \qquad (u-u_h,\varphi) = 0 \quad \forall\ \varphi \in S^h(\Omega).$$

Maximum norm estimates for the L_2 projection were proved by several authors. The approach we shall present here was given in Nitsche [1977].

8A. A Statement of the Main Result; The Connection with Certain Weighted Norms.

Let us first state the main result.

Theorem 8.1. Let $u \in C(\bar{\Omega})$ and $u_h \in S^h(\Omega)$ satisfy (8.1). Then, there exist a constant $c > 0$, independent of u, u_h and h such that

(8.2) $$\|u-u_h\|_{C(\bar{\Omega})} \le c \inf_{\chi\in S^h} \|u-\chi\|_{C(\bar{\Omega})}.$$

Remark 8.1. The inequality (8.2) says that the L_2 projection "behaves somewhat like" a best approximation in maximum norm. It follows immediately from (8.2) and (2.5) that for $u \in C^2(\bar{\Omega})$

(8.3) $$\|u-u_h\|_{C(\bar{\Omega})} \le ch^2 |u|_{C^2(\bar{\Omega})}.$$

Let us first note that in order to prove (8.2) it suffices to prove that P_h is a bounded operator in $C(\bar{\Omega})$, i.e.,

(8.4) $$\|P_h u\|_{C(\bar{\Omega})} \equiv \|u_h\|_{C(\bar{\Omega})} \le C_1 \|u\|_{C(\bar{\Omega})}.$$

where C_1 is independent of μ, u_h and h. This follows easily, for by the triangle inequality

$$\|u-u_h\|_{C(\bar{\Omega})} \le \|u\|_{C(\bar{\Omega})} + \|u_h\|_{C(\bar{\Omega})} \le (1+C_1)\|u\|_{C(\bar{\Omega})}.$$

Now $P\chi = \chi$ for any $\chi \in S^h$ hence

$$\|u-u_h\|_{C(\bar{\Omega})} = \|u-\chi-(u_h-\chi)\|_{C(\bar{\Omega})} \le (1+C_1)\|u-\chi\|_{C(\bar{\Omega})}.$$

from which (8.2) follows.

We shall next derive some relationships between the maximum norms in (8.3) and certain weighted L_2 norms.

We begin with the maximum norm of functions in $S^h(\Omega)$. Let $v \in S^h(\Omega)$, then

$$\|v\|_{C(\overline{\Omega})} = |v(x_0)|$$

for some $x_0 \in \overline{\Omega}$. Let $\tau \in \{\tau\}^h$ be a triangle such that $x_0 \in \overline{\tau}$. By the inverse property (2.16)

$$\|v\|_{C(\overline{\tau})} \leq ch^{-1}\|v\|_{\tau}.$$

Hence

$$\|v\|_{C(\overline{\Omega})} \leq ch^{-1}\left[\int_{\tau} v^2 dx\right]^{1/2}.$$

For any fixed constants $\gamma > 0$ and $C^* \geq 1$, it easily follows from this that since $\operatorname{diam}(\tau) \leq h$

$$(8.4) \qquad \|v\|_{C(\overline{\Omega})} \leq C(1+(C^*)^2)^{\gamma/2}\, h^{-1+\gamma}\left[\int(\rho^2+(C^*h)^2)^{-\gamma}v^2 dx\right]^{1/2},$$

where $\rho = \rho(x) = |x-x_0|$.

For any positive number C^* we set

$$(8.5) \qquad \mu(x) = \rho^2+(C^*h)^2$$

and for any real number γ we defined the weighted norm

$$(8.6) \qquad |||v|||_{\gamma} = \left[\int_{\Omega} \mu^{-\gamma}v^2 dx\right]^{1/2}.$$

Thus for $\gamma > 0$ and $v \in S^h(\Omega)$, (8.4) may be rewritten as

$$(8.7) \qquad \|v\|_{C(\Omega)} \le C(1+(C^*)^2)^{\gamma/2}\, h^{\gamma-1}\, |||v|||_\gamma .$$

On the other hand if $v \in C(\bar{\Omega})$, $C^* \ge 1$ and $\gamma > 1$

$$|||v|||_\gamma = \left[\int_\Omega (\rho^2+(C^*h)^2)^{-\gamma} v^2 dx\right]^{1/2}$$

$$\le \frac{c}{\gamma-1}\, h^{(1-\gamma)} \|v\|_{C(\bar{\Omega})} .$$

Or

$$(8.8) \qquad \frac{\gamma-1}{C}\, h^{\gamma-1}\, |||v|||_\gamma \le \|v\|_{C(\bar{\Omega})} .$$

We next claim that in order to prove (8.3) and hence Theorem 8.1, it (more than) suffices to prove the following:

<u>Theorem 8.2</u>.

$$(8.9) \qquad |||u_h|||_\gamma \le C|||u|||_\gamma ,$$

where C is independent of u_h, u and h.

<u>Remark 8.2</u>. The inequality (8.9) states the P is a bounded operator in the above weighted norms.

Let us first show how (8.3) follows from (8.9). In fact from (8.7), (8.8) and (8.9)

$$\|u_h\|_{C(\bar\Omega)} \le ch^{\gamma-1}(1+(C^*)^2)^{\gamma/2} |||u_h|||_\gamma$$

$$\le ch^{\gamma-1}(1+(C^*)^2)^{\gamma/2} |||u|||_\gamma$$

$$\le \frac{c}{\gamma-1}(1+(C^*)^2)^{\gamma/2}\|u\|_{C(\bar\Omega)}$$

where $\gamma > 1$, which was to be proved.

We now turn to a proof of Theorem 8.2.

8B. A Proof of Theorem 8.1 and 8.2.

For any $\psi \in S^h(\Omega)$

$$(8.10) \qquad |||u_h|||_\gamma^2 = \int_\Omega u_h(\mu^{-\gamma}u_h-\psi)dx + \int_\Omega u_h\psi dx$$

$$= I_1+I_2.$$

In particular we take ψ to be the best approximation to $\mu^{-\gamma}u_h$ in the norm $|||\cdot|||_{-\gamma}$. Thus (see Lemma 3.1)

$$(8.11) \qquad |||\mu^{-\gamma}u_h-\psi|||_{-\gamma} = \inf_{\chi\in S^h} |||\mu^{-\gamma}u_h-\chi|||_{-\gamma}.$$

and

$$(8.12) \qquad |||\psi|||_{-\gamma} \le |||\mu^{-\gamma}u_h|||_{-\gamma} \equiv |||u_h|||_\gamma.$$

We begin by estimating I_2 in (8.10). From (8.1), Schwarz's inequality and (8.12) imply that

$$I_2 = \int_\Omega u_h \psi dx = \int_\Omega u\psi dx = \int_\Omega \mu^{-\gamma/2} u \, \mu^{\gamma/2} \psi dx$$

$$\le |||u|||_\gamma |||\psi|||_{-\gamma}$$

$$\le |||u|||_\gamma |||\mu^{-\gamma} u_h|||_{-\gamma} = |||u|||_\gamma |||u_h|||_\gamma .$$

Hence by the arithmetic-geometric mean inequality

$$I_2 \le 4|||u|||_\gamma^2 + \tfrac{1}{4}|||u_h|||_\gamma^2. \tag{8.13}$$

In view of (8.11)

$$I_1 \le |||u_h|||_\gamma |||\mu^{-\gamma} u_h - \psi|||_{-\gamma} \tag{8.14}$$

$$\le |||u_h|||_\gamma |||\mu^{-\gamma} u_h - (\mu^{-\gamma} u_h)_I|||_{-\gamma}.$$

The function $\mu^{-\gamma} u_h$ has the form of a smooth function $\mu^{-\gamma}$ multiplied by a function $u_h \in S^h$. We shall show in a little while that the following super-approximation result holds:

$$|||\mu^{-\gamma} u_h - (\mu^{-\gamma} u_h)_I|||_{-\gamma} \le C(h|||\mu^{-\gamma-\frac{1}{2}} u_h|||_{-\gamma} + h^2 |||\mu^{-\gamma-1} u_h|||_{-\gamma}). \tag{8.15}$$

Assuming (8.15) for the moment, let us complete the proof of (8.9). Note that

$$\mu^{-1}(x) \lesssim \frac{1}{(C^*h)^2} .$$

Hence from (8.15)

$$|||\mu^{-\gamma}u_h-(\mu^{-\gamma}u_h)_I|||_{-\gamma} \lesssim C\left[\frac{1}{C^*}|||\mu^{-\gamma}u_h|||_{-\gamma}\right]$$

$$\lesssim \frac{C}{C^*}|||u_h|||_{\gamma}.$$

Using this inequality in (8.14) and then the inequalities for I_1 and I_2 in (8.10) we obtain

$$|||u_h|||_{\gamma} \lesssim 4|||u|||_{\gamma}^2 + (\tfrac{1}{4} + \tfrac{C}{C^*})|||u_h|||_{\gamma}^2.$$

where again C is independent of h, u_h, u and $C^* \geq 1$. The inequality (8.9) now trivially follows by taking C^* sufficiently large. We now turn to a proof of (8.15).

<u>8C. Some Properties of S^h in Weighted Norms</u>.

Our object here is to prove (8.15). In doing so we shall implicitly show that approximation results analogous to (2.4) and (2.5) hold as well as the inverse property (2.5) in weighted norms.

<u>Lemma 8.1</u>. Let $\tau \in \{\tau\}^h$.

i) (Approximation) If $u \in H^2(\tau)$ then

$$(8.16)\qquad \int_\tau \mu^{-\gamma}(u-u_I)^2dx \le ch^4 \sum_{|\alpha|=2} \int_\tau \mu^{-\gamma}|D^\alpha u|^2dx.$$

ii) (Inverse property) If $v \in S^h$,

$$(8.17)\qquad \int_\tau \mu^{-\gamma}(\nabla v)^2dx \le ch^{-2} \int_\tau \mu^{-\gamma}v^2dx.$$

iii) (Super-approximation) If $t \neq 0$ and $v \in S^h$,

then

$$(8.18)\qquad \int_\tau \mu^{\gamma}(\mu^t v-(\mu^t w)_I)^2dx$$

$$\le C\{h^2 \int_\tau \mu^\gamma(\mu^{t-1}v)^2dx + h \int \mu^\gamma(\mu^{t-1/2}v)^2dx\}.$$

Furthermore by summing these inequalities over all $\tau \in \{\tau\}^h$ we obtain

iv) If $u \in H^2(\Omega)$, $\gamma \in \mathbb{R}$

$$(8.19)\qquad |||u-u_I|||_\gamma \le ch^2 \sum_{|\alpha|=2} |||D^\alpha u|||_\gamma.$$

v) If $v \in S^h(\Omega)$

$$(8.20)\qquad |||\nabla v|||_\gamma \le ch^{-1}|||v|||_\gamma.$$

vi) If $t \neq 0$, then

$$(8.21)\qquad |||\mu^t v-(\mu^t v)_I|||_\gamma \le C(h|||\mu^{t-1/2}v|||_\gamma + h^2|||\mu^{t-1}v|||_\gamma).$$

<u>Proof</u>: Let us first note that for any $t \in \mathbb{R}$

$$(8.22) \qquad \frac{\max\limits_{x \in \bar{\tau}} (\mu(x))^t}{\min\limits_{x \in \bar{\tau}} (\mu(x))^t} \le 3^{|t|}.$$

We leave the proof of this to the reader. For any $\tau \in \{\tau\}^h$ and any $\gamma \in \mathbb{R}$

$$\int_\tau \mu^{-\gamma}(u-u_I)^2 dx \le (\max_{x \in \bar{\tau}} \mu^{-\gamma}) \int_\tau (u-u_I)^2 dx.$$

Using (2.2) we have

$$\int_\tau \mu^{-\gamma}(u-u_I)^2 dx \le C(\max_{x \in \bar{\tau}} \mu^{-\gamma}) h^4 \sum_{|\alpha|=2} \int_\tau |D^\alpha u|^2 dx$$

$$\le C \frac{(\max\limits_{x \in \bar{\tau}} \mu^{-\gamma})}{(\min\limits_{x \in \bar{\tau}} \mu^{-\gamma})} h^4 \sum_{|\alpha|=2} \int_\tau \mu^{-\gamma} |D^\alpha u|^2 dx,$$

which in view of (8.22) proves (8.16).

The proof of (8.17) follows in a like manner using (2.15).

If we apply (8.16) with $u = \mu^t v$ where $v \in S^h$, it follows that (as in the proof of super-approximation)

$$\int_T \mu^\gamma (\mu^t v - (\mu^t v))^2 dx \le ch^4 \Big\{ \sum_{|\alpha|=2} \int_T \mu^\gamma (D^\alpha \mu^t)^2 v^2 dx$$

$$+ \sum_{|\alpha|,|\beta|=1} \int_T \mu^\gamma (D^\alpha \mu^t)^2 (D^\beta v)^2 dx \Big\}.$$

It is easy to check that for $t \neq 0$

$$|D^\alpha \mu^t| \le \mu^{t-|\alpha|/2}.$$

The inequality (8.18) now follows on combining the last two inequalities with (8.17).

9. Error Estimates in the Maximum Norm for the Ritz Projection.

As for the the L^2 projection, there are several methods which have been developed for proving sharp maximum norm error estimates for the finite element method for elliptic boundary value problems. We shall first state the result which we are aiming for and then briefly discuss the ideas behind two of these methods. A modification of one of them will then be used to prove the result. Again for simplicity, we shall limit ourselves here to studying Dirichlet's Problem I (Section 4) on a convex polygonal domain. Generalizations will be discussed later on. The main result of this section is as follows.

Theorem 9.1. Let Ω be a convex polygonal domain in $\mathbb{R}^2$. Suppose that $u \in C^1(\bar{\Omega}) \cap C_0(\bar{\Omega})$ and $u_h \in S_0^h(\Omega)$ satisfy

$$(9.1) \qquad (\nabla u - \nabla u_h, \nabla \varphi)_\Omega = 0, \quad \text{for all} \quad \varphi \in S_0^h(\Omega).$$

Then there exists a constant C independent of u, u_h and h such that for h sufficiently small

$$(9.2) \qquad \|u-u_h\|_{C(\bar{\Omega})} \leq Ch(\ell n\ 1/h) \inf_{\chi \in S_0^h} \|\nabla(u-\chi)\|_{L^\infty(\Omega)}.$$

Remark 9.1. If $u \in C^2(\bar{\Omega}) \cap C_0(\bar{\Omega})$ then using the approximation result (2.5) together with (9.2) it follows immediately that

$$(9.3) \qquad \|u-u_h\|_{C(\bar{\Omega})} \leq Ch^2(\ell n\ 1/h)|u|_{C^2(\bar{\Omega})}.$$

This is almost optimal from the point of view of approximation theory. Let us also remark that $C^2(\bar{\Omega})$ may be replaced by $W^{2,\infty}(\Omega)$ in (9.3).

Method 1. Estimates Using the Green's Function (An Outline).

Let x be any point of Ω and denote by $G(x,y)$ Green's function for $-\Delta$, i.e., $G(x,y)$ satisfies

$$(9.4)\qquad -\Delta_y G(x,y) = \delta(x-y) \quad \text{in} \quad \Omega$$

and

$$(9.5)\qquad G(x,y) = 0, \ \forall \ y \in \partial\Omega.$$

Here $\delta(x-y)$ is the Dirac measure concentrated at x. In term of G we have the representation

$$(9.6)\qquad (u-u_h)(x) = \int_\Omega \nabla(u(y)-u_h(y))\cdot\nabla G(x,y)dy.$$

Now let $G_h(x,y) \in S_0^h(\Omega)$ be the finite element approximation to the Green's function defined by

$$(9.7)\qquad \int_\Omega \nabla G_h(x,y)\nabla\chi(y)dy = \chi(x), \ \text{for all} \ \ \chi \in S_0^h$$

or equivalently,

$$(9.8)\ \int_\Omega \nabla(G(x,y)-G_h(x,y))\cdot\nabla\chi(y)dy = 0, \ \text{for all} \ \chi \in S_0^h(\Omega).$$

It follows from (9.1) and (9.8) that

$$(9.9)\quad (u-u_h)(x) = \int_\Omega \nabla(u(y)-u_h(y))\cdot\nabla(G(x,y)-G_h(x,y))dy$$

$$= \int_\Omega \nabla(u(y)-\chi(y))\cdot\nabla(G(x,y)-G_h(x,y))dy.$$

This may be estimated by

$$(9.10)\quad |(u-u_h)(x)| \le \|\nabla(u-\chi)\|_{L^\infty(\Omega)} \|\nabla(G(x,\cdot)-G_h(x,\cdot))\|_{L^1(\Omega)}.$$

The estimate (9.2) will follow from (9.10) if we can show that

$$(9.11)\quad \|\nabla(G(x,\cdot)-G_h(x,\cdot)\|_{L^1(\Omega)} \le Ch(\ell n\ 1/h).$$

As estimate of this type is plausible since the second derivatives of $G(x,y)$ "almost" belongs to $L^1(\Omega)$.

Obtaining the estimate (9.11) is the heart of the matter.

This approach was taken by Scott [1976] who first proved (9.11) for the finite element method with quasi-uniform triangulations on a smooth domain. A similar approach had been used for finite difference methods (see for example Bramble [1970]). Scott estimates the left hand side of (9.11) in two parts, i.e.,

$$(9.12) \qquad \|\nabla(G(x,\cdot)-G_h(x,\cdot)\|_{L^1(\Omega)}$$

$$\leq \|\nabla(G(x,\cdot)-G_h(x,\cdot))\|_{L^1(B'(x,C^*h))}$$

$$+ \|\nabla(G(x,\cdot)-G_h(x,\cdot))\|_{L^1(\Omega\setminus B'(x,C^*h))},$$

where $B'(x,C^*h)$ is the intersection of a ball of radius C^*h (C^* chosen appropriately large) with Ω. Note that G is harmonic outside of $B'(x,C^*h)$, which contains the singular part of G. Let us only briefly discuss the second term on the right of (9.12). This may be converted to a weighted L^2 norm by using the Schwarz inequality, i.e.,

$$\int_{\Omega\setminus B'} |\nabla(G(x,y)-G_h(x,y))|dy = \int_{\Omega\setminus B'} \rho^{-1}\rho|\nabla(G-G_h)|dy$$

$$\leq \left[\int_{\Omega\setminus B'} \rho^2|\nabla(G-G_h)|^2dy\right]^{1/2}\left[\int_{\Omega\setminus B'} \rho^{-2}dy\right]^{1/2}$$

$$\leq C(\ell n\,\tfrac{1}{h})^{1/2}\left[\int_{\Omega\setminus B'} \rho^2|\nabla G-G_h|^2dy\right]^{1/2}.$$

This last weighted L^2 integral can be estimated using the ideas used in the proof of Theorem 6.1, where essentially the "cutoff" function w is replaced by the non-vanishing weight ρ^2 on $\Omega\setminus B'$ and $\rho = |x-y|$.

<u>Method 2. Estimates Using Weighted Norms</u>. This method was first used by Natterer [1975] and then sharpened, generalized and put into its present elegant form by Nitsche [1975]. Let

x_0 be the point in $\bar{\Omega}$ where $\|u_h\|_{C(\bar{\Omega})} = |u_h(x_0)|$, and let $x_0 \in \tau$ for some triangle $\tau \in \tau^h$. In Section 8 we showed that

$$\|u_h\|_{C(\bar{\Omega})} \le Ch^{-1}\|u_h\| \le Ch^{-1+\gamma}\left[\int_\Omega (\rho^2+C^*h)^{-\gamma}u_h^2dx\right]^{1/2}, \tag{9.14}$$

for C^* a constant to be appropriately chosen and where $\rho = |x-x_0|$. Furthermore

$$h^{-1+\gamma}\left[\int_\Omega (r^2+(C^*h)^2)^{-\gamma}u^2dx\right]^{1/2} \le C\|u\|_{C(\bar{\Omega})} \quad \text{for} \quad \gamma > 1. \tag{9.15}$$

The last two inequalities say that the weighted L^2 norm shown above is equivalent to the maximum norm on the subspace $S_0^h(\Omega)$ provided $\gamma > 1$. As in Section 8, for $C^* = 1$ a positive constant to be chosen we set

$$|||u|||_\gamma = \left[\int_\Omega (\rho^2+(C^*h)^2)^{-\gamma}\ u^2dx\right]. \tag{9.16}$$

Nitsche [8] proved a general result for domains with a smooth boundary which when restricted to 2 dimensions takes the form

$$|||u_h|||_{\gamma+1} + |||\nabla u_h|||_\gamma \le C(|||u|||_{\gamma+1}+|||\nabla u|||_\gamma), \tag{9.17}$$

for $\alpha \in (1,2)$. This says that the finite element projection is bounded in the above weighted norms and the appropriate error estimates follow from (9.14)-(9.17) and a weighted norm approximation results.

Let us finish the discussion of this method by indicating how the proof of (9.17) starts.

Let $\mu = (\rho^2+(C^*h)^2)$, then for the second term on the left of (9.17) we have

$$(9.18)\quad |||\nabla u_h|||_\gamma^2 = (\mu^{-\gamma}\nabla u_h, \nabla u_h)_\Omega = (\nabla u_h, \nabla(\mu^{-\gamma}u_h))_\Omega - (\nabla u_h, u_h\nabla\mu^{-\gamma})_\Omega$$

$$= (\nabla u_h, \nabla(\mu^{-\gamma}u_h))_\Omega - \tfrac{1}{2}(\nabla(u_h)^2, \nabla\mu^{-\gamma})_\Omega$$

$$= (\nabla u_h, \nabla(\mu^{-\gamma}u_h))_\Omega + \tfrac{1}{2}(u_h^2, \nabla\mu^{-\gamma})_\Omega$$

$$= I_1 + I_2.$$

For I_2, we have that since $\nabla(\mu^{-\gamma}) \lesssim C\mu^{-\gamma-1}$

$$(9.19)\qquad |I_2| \le C|||u_h|||_{\gamma+1}^2.$$

In view of (9.1) we have, for I_1 the identity

$$(9.20)\quad (\nabla u_h, \nabla(\mu^{-\alpha}u_h))_\Omega = (\nabla u_h, \nabla(\mu^{-\gamma}u_h - \chi))_\Omega - (\nabla u, \nabla(\mu^{-\gamma}u_h - \chi))_\Omega + (\nabla u, \nabla\mu^{-\gamma}u_h)_\Omega$$

for any $\chi \in S_0^h(\Omega)$.

For the first term on the right of (9.20)

$$|(\nabla u_h, \nabla(\mu^{-\alpha} u_h - \chi))_\Omega| = |(\mu^{-\gamma/2} \nabla u_h, \mu^{\gamma/2} \nabla(\mu^{-\alpha} u_h - \chi))_\Omega|$$

$$\leq |||\nabla u_h|||_\gamma \, |||\nabla(\mu^{-\gamma} u_h - \chi)|||_{-\gamma}.$$

Notice that as for the L_2 projection, $\mu^{-\gamma} u_h$ is of the form of a smooth function multiplied by a function in the subspace. One can then use the super-approximation property given in Section 8. We shall not give further details here but instead refer the reader to Nitsche [1975] where a generalization of Theorem 9.1 is proved on smooth domains. Let us just remark that the first step (9.18) in the proof is directly analogous to the proof of local estimates with the cut off function ω now replaced by the weight $\mu^{-\gamma}$.

We now turn to a complete proof of Theorem 6.1 (see Schatz and Wahlbin [1978]).

Step 1: We shall first reduce the proof of (9.2) to the problem of finding an estimate for the Galerkin projection of a function v which may be thought of as a smoothed out and renormalized Green's function. This will be done by first establishing a simple relationship between the maximum norm of the error and the L^2 norm on some triangle T. The L^2 norm of the error will be estimated by a duality argument, where v is the solution of the dual problem.

Let $x_0 \in \bar{\Omega}$ be such that

$$\|u - u_h\|_{C(\bar{\Omega})} = |(u - u_h)(x_0)|. \tag{9.21}$$

Now $x_0 \in T$ for some triangle T. Let u_I be the linear interpolant of u on $\overline{T}$, then successively using the triangle inequality, the inverse property (3.20) and the triangle inequality once more, it follows that

$$|(u-u_h)(x_0)| \leq |(u(x_0)-u_I(x_0)|+|(u_I-u_h)(x_0)|$$

$$\leq |u(x_0)-u_I(x_0)|+Ch^{-1}\|u_I-u_h\|$$

$$\leq |u(x_0)-u_I(x_0)|+Ch^{-1}\|u-u_I\|_T+Ch^{-1}\|u-u_h\|_T.$$

Since the area$(T) \leq Ch^2$, $\|(u-u_I)\|_{L^2(T)} \leq Ch\|u-u_I\|_{C(\overline{T})}$ and hence

$$\|u-u_h\|_{C(\overline{\Omega})} \leq C(\|u-u_I\|_{C(\overline{T})}+h^{-1}\|u-u_h\|_T).$$

Using the approximation result (2.5), it then follows that

$$\|u-u_h\|_{C(\overline{\Omega})} \leq Ch\|\nabla u\|_{L_\infty(\overline{\Omega})}+Ch^{-1}\|u-u_h\|_T. \tag{9.22}$$

The second term on the right will be estimated by a duality argument. Now

$$\|u-u_h\|_T = (u-u_h,\psi)_\Omega, \tag{9.23}$$

where ψ is defined by

$$\psi(x) = \begin{cases} (u-u_h)(x)/\|u-u_h\|_T & \text{for} \quad x \in \overline{T} \\ 0 & \text{for} \quad x \in \overline{\Omega}-\overline{T} \end{cases}.$$

Notice that $\psi \in L^2(\Omega)$ with $\|\psi\|_\Omega = \|\psi\|_T = 1$. Let $v \in H_0^1(\Omega)$ be the unique solution of the boundary value problem

$$(9.24) \qquad (\nabla v, \nabla\varphi)_\Omega = (\psi, \varphi)_\Omega, \text{ for all } \varphi \in H_0^1(\Omega).$$

From Theorem 4.1 we have that $v \in H^2(\Omega) \cap H_0^1(\Omega)$ and

$$(9.24a) \qquad \|v\|_{2,\Omega} \leq C.$$

Notice that v is harmonic on $\Omega-\overline{T}$ and may be thought of as a smoothed out (and renormalized) Green's function. Using (9.24) with $\varphi = u-u_h$ in (9.23), we obtain

$$(9.25) \qquad \|u-u_h\|_T = (\nabla(u-u_h), \nabla v)_\Omega.$$

Now let $v_h \in S_0^h(\Omega)$ be the Galerkin finite element approximation of v defined by

$$(9.26) \qquad (\nabla(v-v_h), \nabla\varphi)_\Omega = 0, \text{ for all } \varphi \in S_0^h.$$

Then

$$(9.27)\qquad \|u-u_h\|_T = (\nabla(u-u_h),\nabla(v-v_h))_\Omega = (\nabla u,\nabla(v-v_h))_\Omega$$

$$\le \|\nabla u\|_{L^\infty(\Omega)} \|\nabla(v-v_h)\|_{L^1(\Omega)} .$$

Combining (9.27) and (9.22) we arrive at

$$(9.28)\qquad \|u-u_h\|_{C^\infty(\bar\Omega_0)} \le Ch\|\nabla u\|_{L^\infty(\Omega)} \left[1+h^{-2}\|\nabla(v-v_h)\|_{L^1(\Omega)}\right],$$

and the proof of Theorem 9.1 will be complete once the following has been proved:

Step 2.

Lemma 9.1. There exists a constant C independent of h, v, v_h, and x_0 such that

$$(9.29)\qquad \|\nabla(v-v_h)\|_{L^1(\Omega)} \le Ch^2\ell n(1/h).$$

Proof: Let us first remark that the estimate (9.29) is analogous to (9.11), except that the function v treated here is normalized in a different way. For technical reasons we shall need to choose a fixed positive constant $\hat{C}$ (independent of h and x_0) in the following way: Let C^* be as in Theorem 9.1 and C^{**} be such that $\max \operatorname{diam}(\tau) \le \frac{C^{**}}{h} h$. We then choose $\hat{C} = \max(8C^*, 8C^{**})$. Having fixed $\hat{C}$ we turn to proving (9.29). Now let $\hat{\Omega} = B(x_0, \hat{C}h) \cap \Omega$ then (analogous to (9.12))

$$(9.30)\qquad \|\nabla(v-v_h)\|_{L^1(\Omega)} = \|\nabla(v-v_h)\|_{L^1(\hat{\Omega})} + \|\nabla(v-v_h)\|_{L^1(\Omega/\hat{\Omega})}$$

$$= J_1+J_2.$$

Using the Cauchy-Schwarz inequality, Theorem 4.2 and (9.24)

$$(9.31)\qquad |J_1| \le Ch\|\nabla(v-v_h)\|_{\hat{\Omega}} \le Ch\|\nabla(v-v_h)\|_{\Omega}$$

$$\le Ch^2|v|_{2,\Omega} \le Ch^2.$$

In order to estimate J_2 we decompose the domain $\Omega\setminus\hat{\Omega}$ as follows: Let A_j be the annular regions

$$A_j = \{x : 2^{-j-1} \le |x-x_0| \le 2^{-j}\},\ j = 0,1,2,\ldots \text{ etc.}$$

Set

$$\Omega_j = A_j \cap \Omega$$

and

$$d_j = 2^{-j}.$$

Without loss of generality we may assume that $\text{diam}(\Omega) \le 1$, then $\overline{\Omega} \subseteq \bigcup_{j=0}^{\infty} \Omega_j$. Let J be defined by

$$2^{-J-1} \le \hat{C}h \le 2^{-J}.$$

Notice that $J = O(\ell n \frac{1}{h})$ and $\overline{\Omega/\hat{\Omega}} \subseteq \bigcup_{j=0}^{J} \Omega_j$. We shall need some further notation. Let

$$\Omega_j' = \Omega_{j-1} \cup \Omega_j \cup \Omega_{j+1}$$
$$\Omega_j'' = \Omega_{j-1}' \cup \Omega_j' \cup \Omega_{j+1}' \text{ . etc.}$$

Now using the Cauchy-Schwarz inequality

$$J_2 \le \sum_{j=0}^{J} \|\nabla(v-v_h)\|_{\Omega_j} \le \sum_{j=0}^{J} d_j \|\nabla(v-v_h)\|_{\Omega_j} .$$

Note that $d_j \ge C^* h$ in this sum. Applying Corollary 6.1 of Theorem 6.1 to the regions Ω_j and Ω_j' respectively it follows from (6.1), with χ chosen to be the interpolant of V that

$$J_2 \le C\Big[\sum_{j=0}^{J} d_j h|v|_{2,\Omega_j'} + \sum_{j=0}^{J} \|v-v_h\|_{\Omega_j'}\Big]. \tag{9.32}$$

For the second term on the right of (9.32)

$$(9.33)\qquad \sum_{j=0}^{J} \|v-v_h\|_{\Omega_j'} \le (\ell n\ 1/h)^{1/2}\Big[\sum_{j=0}^{J} \|v-v_h\|^2_{\Omega_j'}\Big]^{1/2}$$

$$\le (\ell n\ 1/h)^{1/2}\|v-v_h\|_{\Omega}$$

$$\le Ch^2(\ell n\ 1/h)^{1/2}|v|_{2,2,\Omega}$$

$$\le Ch^2(\ell n\ 1/h)^{1/2}.$$

Here we have used the Cauchy-Schwarz inequality, Theorem 4.3 and (9.24a). Combining (9.33), (9.32) and (9.31) with (9.30) we arrive at

$$(9.34)\qquad \|\nabla(v-v_h)\|_{L^1(\Omega)} \le C\Big[h^2(\ell n\ 1/h)^{1/2} + h\sum_{j=0}^{J} d_j|v|_{2,\Omega_j'}\Big].$$

Because of our choice of $\hat{C}$, the function v is harmonic inside Ω_j'', $j = 0,\dots,J$ and the following local estimate holds.

$$(9.35)\qquad |v|_{2,\Omega_j'} \le Cd_j^{-1}\|\nabla v\|_{\Omega_j''}\ .$$

Granting this estimate for a moment, let us complete the proof of (9.29). Now for the second term on the right in (9.34)

$$(9.36)\qquad h\sum_{j=0}^{J} d_j|v|_{2,\Omega_j'} \le Ch\sum_{j=0}^{J}\|\nabla v\|_{\Omega_j''}$$

$$\le Ch(\ell n\ 1/h)^{1/2}\Big[\sum_{j=0}^{J}\|\nabla v\|^2_{\Omega_j''}\Big]^{1/2}$$

$$\le Ch(\ell n\ 1/h)^{1/2}\|\nabla v\|_{\Omega}.$$

In order to estimate this last term we use

$$\|\nabla v\|^2 = (\nabla v,\nabla v)_\Omega = (\psi,v)_\Omega = (\psi,v-v_h)_\Omega+(\psi,v_h)_\Omega$$

$$\le \|\psi\|_\Omega\|v-v_h\|_\Omega+\|\psi\|_{L^1(T)}\|v_h\|_{C(\bar{T})}.$$

Using (9.24), Theorem 4.3, the estimate $\|\psi\|_{L^1(T)} \le Ch\|\psi\|_T = Ch$, and the Sobolev inequality (3.23)

$$\|\nabla v\|^2_\Omega \le Ch(\|\nabla v\|_\Omega+(\ell n\ 1/h)^{1/2}\|\nabla v_h\|_\Omega$$

$$\le Ch(1+(\ell n\ \tfrac{1}{h})^{1/2})\|\nabla v\|_\Omega$$

or

$$\|\nabla v\|_\Omega \le Ch(\ell n\ 1/h)^{1/2}.$$

This estimate combined with (9.36) and (9.34) proves (9.29). We now turn to the proof of (9.35).

Let $w \in C_0^\infty(\Omega_j'')$, $w \geq 0$, $w \equiv 1$ on Ω_j and

$$|D^\alpha w| \leq Cd^{-|\alpha|}, \quad |\alpha| = 1,2.$$

We consider two cases. If Ω_j'' does not intersect $\partial\Omega$ then for any constant M, $w(v-M) \in W^{2,2}(\Omega) \cap W_0^{1,2}(\Omega)$. Using the a priori estimate (2.3) and the fact that $\Delta(v-M) = 0$ on Ω_j'',

$$(9.37) \qquad |v|_{2,\Omega_j'} = |v-M|_{2,\Omega_j'} \leq |w(v-M)|_{2,\Omega} \leq \|\Delta w(v-M)\|_\Omega$$

$$= \|w\Delta(v-M)+\nabla w\cdot\nabla(v-M)+(v-M)\Delta w\|_\Omega$$

$$\leq C(d_j^{-1}\|\nabla v\|_{\Omega_j''}+d_j^{-2}\|v-M\|_{\Omega_j''}).$$

Choosing $M = \int_{\Omega_j''} v\,dx$ it easily follows from a Poincaré inequality that

$$\|v-M\|_{\Omega_j''} \leq d_j\|\nabla(v-M)\|_{\Omega_j''} = d_j\|\nabla v\|_{\Omega_j''} .$$

This combined with (9.37) yield (9.35). If $\Omega_j'' \cap \partial\Omega \neq 0$ we take $M = 0$ and from (3.77) obtain

$$|v|_{2,\Omega_j'} \leq C(d_j^{-1}\|\nabla v\|_{\Omega_j''}+d_j^{-2}\|v\|_{\Omega_j''}).$$

Extending v to be zero outside of Ω, then

$$\|v\|_{\Omega_j''} = \|v\|_{A_j''} \le \|v\|_{A_j'''} \, .$$

Now A_j''' contains a set of measure $\ge Cd^2$ on which v vanishes and a Poincaré inequality yields

$$\|v\|_{A_j'''} \le d_j \|\nabla v\|_{A_j'''} = d_j \|\nabla v\|_{\Omega_j'''} \, .$$

Taken together these last three inequalities yield (9.35), which completes the proof.

Let us end this section with a discussion of other results relating to maximum norm estimates.

First let Ω be a bounded domain in $\mathbb{R}^N$, $N \ge 2$ (for the case $N = 1$ see Douglas, Dupont and Wahlbin [1975a]) with a smooth boundary. Let $S_0^{h,r}$ denote the space of piecewise polynomials of degree $r-1$, $r \ge 2$ on a quasi-uniform triangulation of Ω which vanish on $\partial\Omega$ (i.e., the boundary is assumed to be fitted exactly). Using Method 2, Nitsche proved

$$\|u-u_h\|_{C(\overline{\Omega})} \le Ch(\ell n \tfrac{1}{h})^{\overline{r}} \inf_{\chi \in S_0^{h,r}} \|\nabla(u-\chi)\|_{L_\infty(\overline{\Omega})} \, ,$$

where $\overline{r} = 1$ if $r = 2$, and $\overline{r} = 0$ if $r \ge 3$. Using isoparametric elements Schatz and Wahlbin [1982] showed

(9.38) $$\|u-u_h\|_{C(\overline{\Omega})} \le C(\ell n \tfrac{1}{h})^{\overline{r}} \inf_{\chi \in S_0^{h,r}} \|u-\chi\|_{C(\overline{\Omega})} \, .$$

which says the Galerkin finite element projection is maximum norm stable for $r \geq 3$ and almost maximum norm stable for $r = 2$. In Schatz [1980] it was shown that (9.38) also holds on plane polygonal domains (both convex and non-convex). In the case that Ω is a plane convex polygonal domain Rannacher and Scott [1982] have proved

$$\|\nabla(u-u_h)\|_{L_\infty} \leq C \inf_{\chi \in S_0^h} \|\nabla(u-\chi)\|_{L_\infty} .$$

which shows that the gradients of the Galerkin finite element projection are stable in L^∞ for piecewise linear functions. For $S_0^{h,r}$, $r \geq 3$ this result was known previously, and for S_0^h this nicely removes a logarithm from known estimates. For a rather complete bibliography on maximum norm estimates the reader is referred to Nitsche [1977].

References

Adams, R. A. [1975] Sobolev Spaces, Academic Press, New York.

Aubin, J. P. [1967] Behavior of the error in the approximation of boundary value problems for linear elliptic operators by Galerkins and finite difference methods, Scuola Norm. Sup. Pisa, 21, 599-637.

Babuska, I. [1970] Finite element methods for domains with corners, Computing 6, 264-273.

Babuska, I. [1973] The finite element method with Lagrangian multipliers, Numer. Math. 20, 179-192.

Babuska, I. and Aziz, A. K. [1976] On the angle condition in the finite element method, SIAM J. Numer. Anal. 13, 214-226.

Babuska, I., Miller, A. and Vogelius, M. [1983] Adaptive methods and error estimation for elliptic problems of structural mechanics, Tech. Note BN-1009, University of Maryland.

Babuska, I. and Osborn, J. [1980] Analysis of finite element methods for second order boundary value problems using mesh dependent norms, Numer. Math. 34, 41-62.

Babuska, I. and Rosenzweig, M. [1973] A finite element scheme for domains with corners, Numer. Math. 20, 1-21.

Bergh, J. and Löfström, J. [1976] Interpolation Spaces: An Introduction, Springer, New York.

Blair, J. [1970] Approximate solution of elliptic and parabolic boundary value problems, Thesis, University of California, Berkeley.

Bramble, J. H. [1970] On the convergence of difference approximations for second order uniformly elliptic operators, SIAM-AMS Proceedings 2, 1970, 201-210.

Bramble, J. H. [1975] A survey of some finite element methods proposed for treating the Dirichlet problem, Adv. in Math. 16, 187-196.

Bramble, J. H. [1981] The Lagrange multiplier method for Dirichlet's problem, Math. Comp. 37, 1-12.

Bramble, J. H. and Hilbert, S. R. [1970] Estimation of linear functionals on Sobolev spaces with application to Fourier transforms and spline interpolation, SIAM J. Numer. Anal. 7, 113-124.

Bramble, J. H. and Hilbert, S. R. [1971] Bounds for a class of linear functionals with application to Hermite interpolation, Numer. Math. 16, 362-369.

Bramble, J. H., Nitsche, J. A., and Schatz, A. H. [1975] Maximum norm interior estimates for Ritz-Galerkin methods, Math. Comp. 29, 677-688.

Bramble, J. H. and Osborn, J. E. [1973] Rate of convergence estimates for non-selfadjoint eigenvalue approximation, Math. Comp. 27, 525-549.

Bramble, J. H. and Schatz, A. H. [1976] Estimates for spline projections, RAIRO Anal. Num. 10, 5-37.

Bramble, J. H. and Schatz, A. H. [1977] Higher order local accuracy by averaging in the finite element method, Math. Comp. 31, 94-111.

Christie, I., Griffiths, D. F., Mitchell, A. R., and Zienkiewicz, O. C. [1976] Finite element methods for second order differential equations with significant first derivatives, Internat. J. Numer. Meths. Engrg. 10, 1389-1396.

Ciarlet, Ph. [1978] <u>The Finite Element Method for Elliptic Problems</u>, North-Holland, Amsterdam.

Crouzeix, M. and Thomée, V. [1986a] The stability in L_p and W_p^1 of the L_2 projection into finite element function spaces, to appear.

De Boor, C. [1979] On a max-bound for the least-squares spline approximant, in: <u>Approximation and Function Spaces</u>, Z. Ciesielski, Ed., North Holland, Amsterdam, 163-175.

De Boor, C. and Fix, G. J. [1973] Spline approximation by quasiinterpolants, J. Approximation Theory 8, 19-45.

Demko, S. [1977] Inverses of band matrices and local convergence of spline projections, SIAM J. Numer. Anal. 14, 616-619.

Descloux, J. [1972] On finite element matrices, SIAM J. Numer. Anal. 9, 260-265.

Descloux, J. [1976] Interior regularity and local convergence of Galerkin finite element approximations for elliptic equations, Topics in Numerical Analysis II, J. Miller Ed., Academic Press, 1976.

Douglas, J. and Dupont, T. [1974] Galerkin approximation for the two-point boundary problem using continuous piecewise polynomial spaces, Numer. Math. 22, 99-109.

Douglas, J., Dupont, T., and Wahlbin, L. B. [1975a] Optimal L_∞ error estimates for Galerkin approximations to solutions of two-point boundary value problems, Math. Comp. 39, 475-483.

Douglas, J., Dupont, T., and Wahlbin, L. B. [1975b] The stability in L^q of the L^2-projection into finite element function spaces, Numer. Math. 23, 193-197.

Douglas, J., Dupont, T., and Wheeler, M. F. [1974a] An L^∞ estimate and a superconvergence result for a Galerkin method for elliptic equations based on tensor products of piecewise polynomials, RAIRO Anal. Numer. 8, 61-66.

Douglas, J., Dupont, T. and Wheeler, M. F. [1974b] A Galerkin procedure for approximating the flux on the boundary for elliptic and parabolic boundary value problems, RAIRO Anal. Numer. 8, 47-59.

Douglas, J. and Milner, F. [1985] Interior and superconvergence estimates for mixed methods for second order elliptic problems, RAIRO M^2 AN 19(3), 397-428.

Eriksson, K. [1985] Improved accuracy by adapted mesh refinements in the finite element method, Math. Comp. 44, 321-344.

Eriksson, K. and Johnson, C. [1986] An adaptive finite element method for linear elliptic problems, to appear.

Eriksson, K. and Thomée, V. [1984] Galerkin methods for singular boundary value problems in one space dimension, Math. Comp. 42, 345-367.

Frehse, J. and Rannacher, R. [1976] Eine L^1-Fehlerabschätzung diskreter Grundlösungen in der Methode der finiten Elemente, Tagungsband "Finite Elemente", Bonn. Math. Schr.

French, D. [1985] The finite element method for a degenerate elliptic equation, Thesis, Cornell University.

Gallagher, R. H. [1975] Survey and evaluation of the finite element method in linear fracture mechanics analysis, Computational Fracture Mechanics, Second National Congress on Pressure Vessels and Piping, The American Society of Mechanical Engineers, New York, 637-653.

Grisvard, P. [1976] Behavior of the solution of an elliptic boundary value problem in a polygonal or polyhedral domain, in: Numerical Solution of Partial Differential Equations - III, B. Hubbard, Editor, Academic Press, New York, 207-274.

Grisvard, P. [1985a] Boundary Value Problems in Non-Smooth Domains, Pitman Publishing Co., Boston, Mass.

Grisvard, P., Wendland, W. and Whiteman, J. R., Eds., [1985b] Singularities and Constructive Methods for Their Treatment, Springer Lecture Notes in Mathematics 1121.

Haverkamp, R. [1982] Eine Aussage zur L_∞-stabilität und zur genauen Konvergenzordnung der H_0^1-Projection, Preprint No. 508, Universität Bonn.

Hellen, T. K. [1980] Numerical methods in fracture mechanics, Developments in Fracture Mechanics-1, G.C.Chell, Ed., Applied Science Publishers, Barking, England, 145-181.

Hilbert, S. [1973] A mollifier useful for approximations in Sobolev spaces and some applications to approximating solutions of differential equations, Math. Comp. 27, 81-89.

Hughes, T. J. R. and Brooks, A. N. [1979] A multidimensional upwind scheme with no crosswind diffusion, in: Finite Element Methods for Convection Dominated Flows, Hughes, T.J.R. Ed., AMD, 34, ASME, New York, 19-35.

Jamet, P. [1976] Estimations d'erreur pour des elements finis droits preque dégénérés, RAIRO, Anal. Numer. 10, 43-61.

Jespersson, D. [1978] Ritz-Galerkin methods for singular boundary value problems, SIAM J. Numer. Anal. 15, 813-834.

Johnson, C. and Nävert, U. [1981] An analysis of some finite element methods for advection-diffusion problems, in: Analytical and Numerical Approaches to Asymptotic Problems in Analysis, Axelsson, O., Frank, L. S., and Van der Sluis, A., Eds., North Holland, Amsterdam, 99-116.

Kellogg, R. B. [1971] Interpolation between subspaces of a Hilbert space, Technical Note BN-719, University of Maryland.

Kellogg, R. B. [1972] Higher order singularities for interface problems, in: The Mathematical Foundations of the Finite Element Method, A. K. Aziz, Editor, Academic Press, New York, 589-602.

Kondrat'ev, V. A. [1967] Boundary problems for elliptic equations in domains with conical or angular points, Trans. Moscow Math. Soc. v. 16, 227-313.

Laasonen, P. [1967] On the discretization error of the Dirichlet problem in a plane region with corners, Ann. Acad. Sci. Fenn. Ser. A., I Math. v. 408, 1-16.

LeSaint, P. and Zlamal, M. [1979] Superconvergence of the gradient of finite element solutions, RAIRO Anal. Numer. 13, 139-168.

Levine, N. [1985] Superconvergent recovery of the gradient from piecewise linear finite-element approximations, IMA Journal Numer. Anal. 5, 407-427.

Lin Qun, Lu Tao and Shen Shumin [1983] Asymptotic expansion for finite element approximations, Institute of Mathematical Sciences, Chengdu Branch of Academia Sinica.

Lin Qun and Xu Jinchao [1985] Linear finite elements with high accuracy, J. Comp. Math. 3.

Lions, J. L. [1973] Perturbations Singulières dans les Problèmes aux Limites et en Controle Optimal, Lecture Notes in Math. 323, Springer, New York.

Lorentz, G. G. [1966] Approximation of Functions, Holt, Rinehart and Winston, New York.

Natterer, F. [1975] Über die Punktweise Konvergenz Finiter Elemente, Numer. Math. 25, 67-77.

Natterer, F. [1977] Uniform convergence of Galerkin's method for splines on highly nonuniform meshes, Math. Comp. 31, 457-468.

Nävert, U. [1982] A finite element method for convention-diffusion problems, Thesis, Chalmers University of Technology and the University of Gothenburgh.

Nitsche, J. [1968] Ein kriterium für die quasi-optimalität des Ritzschen Verfahrens, Numer. Math. vol. 11, 346-348.

Nitsche, J. A. [1972a] Interior error estimates of projection methods, in: Proceedings of the Conference on Differential Equations and their Applications, EQUADIFF 3, J. E Purkyne University, Brno, pp. 235-239.

Nitsche, J. A. [1972b] On Dirichlet problems using subspaces with nearly zero boundary conditions, in: The Mathematical Foundations of the Finite Element Method, A. K. Aziz, Ed., Academic Press, New York, NY, 603-627.

Nitsche, J. A. [1975] L_∞ convergence for finite element approximation, 2. Conference on Finite Elements, Rennes, France.

Nitsche, J. A. [1976a] Zur lokalen Konvergenz von Projectionen auf finite Elemente, in: Approximation Theory, Lecture Notes in Math., vol. 556, Springer, 329-346.

Nitsche, J. A. [1976b] Der Einfluss von Randsingularitäten beim Ritzcshen Verfahren, Numer. Math. 25, 263-278.

Nitsche, J. A. [1977] L_∞ convergence of finite element approximations, Mathematical Aspects of Finite Element Methods, Lecture Notes in Math. vol. 606, Springer-Verlag, New York, 261-274.

Nitsche, J. A. [1979] Interior error estimates for semidiscrete Galerkin approximations for parabolic equations, Conference on Progress in the Theory and Practice of the Finite Element Method, Chalmers, Götenborg, Sweden.

Nitsche, J. A. and Schatz, A. H. [1972] On local approximation properties of L_2-projection on spline-subspaces, Applic. Anal. 2, 161-168.

Nitsche, J. A. and Schatz, A. H. [1974] Interior estimates for Ritz-Galerkin methods, Math. Comp. 28, 937-958.

Oganesyan, L. A. and Rukhovets, L. A. [1969] Study of the rate of convergence of variational difference schemes for second order elliptic equations in a two-dimensional field with a smooth boundary, USSR Comp. Math. and Math. Physics, 9, 158-183.

Raithby, G. D. and Torrance, K. E. [1974] Upstream-weighted differencing schemes and their application to elliptic problems involving fluid flow, Computer and Fluids 2, 191-206.

Rannacher, R. and Scott. R. [1982] Some optimal error estimates for piecewise linear finite element approximations, Math. Comp. 38, 437-445.

Raugel, G. [1978] Résolution numérique de problèmes elliptiques dans des domaines avec coins, C. R. Acad. Sci. Paris, Ser A-B 286, no. 18, A791-794.

Rice, J. R. [1969] On the convergence of nonlinear spline approximation, in: Approximations with Special Emphasis on Spline Functions, I. J. Schoenberg, Ed., Academic Press, New York, 349-365.

Riemann, B. [1854] Über die Darstellbarkeit einer Funktion durch eine trigonometrische Reihe, Collected Works, 2nd Ed., 227-271, Leipzig 1892.

Schatz, A. H. [1974] An observation concerning Ritz-Galerkin methods with indefinite bilinear forms, Math. Comp. v. 28, 959-962.

Schatz, A. H. [1980] The finite element method on polygonal domains, in: Seminar on Numerical Analysis and its Applications to Continuum Physics, Colecao ATAS, Rio de Janeiro, 57-64.

Schatz, A. H. [1980] A weak discrete maximum principle and stability of the finite element method in L^∞ on plane polygonal domains, Math. Comp. 34, 77-91.

Schatz, A. H. and Wahlbin, L. B. [1977] Interior maximum norm estimates for finite element methods, Math. Comp. 31, 414-442.

Schatz, A. H. and Wahlbin, L. B. [1978] Maximum norm estimates in the finite element method on plane polygonal domains, Part 1, Math. Comp. 32, 73-109.

Schatz, A. H. and Wahlbin, L. B. [1979] Maximum norm estimates in the finite element method on plane polygonal domains, Part 2, Refinements, Math. Comp. 33, 465-492.

Schatz, A. H. and Wahlbin, L. B. [1981] On a local asymptotic error estimate in finite elements and its use: Numerical examples, in: Advances in Computer Methods for Partial Differential Equations IV, R. Vichnevetsky and R. S. Stepleman, Eds., IMACS, 14-17.

Schatz, A. H. and Wahlbin, L. B. [1982] On the quasi-optimality in L_∞ of the $\mathring{H}^1$-projection into finite element spaces, Math. Comp. 38, 1-21.

Schatz, A. H. and Wahlbin, L. B. [1983] On the finite element for singularly perturbed reaction-diffusion problems in two and one dimensions, Math. Comp. 40, 47-89.

Schreiber, R. [1980] Finite element methods of high order accuracy for singular two-point boundary value problems with non-smooth solutions, SIAM J. Numer. Anal. 17, 547-566.

Scott, R. [1973] Finite element convergence for singular data, Numer. Math. 21, 317-327.

Scott, R. [1975] Interpolated boundary conditions in the finite element method, SIAM J. Numer. Anal. 12, 404-427.

Scott, R. [1976] Optimal L^∞ estimates for the finite element method on irregular meshes, Math. Comp. 30, 681-697.

Stein, E. M. [1970] Singular Integrals and Differentiability Properties of Functions, Princeton Univ. Press, Princeton, N.J.

Strang, G. [1973] Approximation in the finite element method, Numer. Math. 19, 81-98.

Strange, G. and Fix, G. J. [1973] An Analysis of the Finite Element Method, Prentice-Hall, Englewood Cliffs, NJ.

Thatcher, R. W. [1976] The use of infinite grid refinements at singularities in the solution of Laplace's equation, Numer. Math. 25, 163-178.

Thomée, V. [1977] Higher order local approximations to derivatives in the finite element method, Math. Comp. 31, 652-660.

Wahlbin, L. B. [1978] Maximum norm error estimates in the finite element method with isoparametric quadratic elements and numerical integration, RAIRO, Anal. Numer. 12, 173-202.

Wahlbin, L. B. [1984] On the sharpness of certain local estimates for $\mathring{H}^1$ projections into finite element spaces: Influence of a reentrant corner, Math. Comp. 42, 1-8.

Wahlbin, L. B. [1985] A comparison of the local behavior of spline L^2 projections, Fourier series and Legendre series, in Singularities and Constructive Methods for Their Treatment (R. Grisvard, W. Wendland and J. R. Whiteman, Eds.), Lecture Notes in Mathematics 1121, Springer, New York.

Wendland, W. L. [1979] Elliptic Systems in the Plane, Pitman, San Francisco, CA.

Zlamal, M. [1977] Some superconvergence results in the finite element method, in: Mathematical Aspects of Finite Element Methods, Lecture Notes in Mathematics 606, Springer, 353-362.

THE FINITE ELEMENT METHOD FOR PARABOLIC PROBLEMS

Vidar Thomée

1. INTRODUCTION.

In this introductory section we shall consider the standard Galerkin method for the approximate solution of the initial-boundary value problem for the heat equation.

Let Ω be a domain in R^d with smooth boundary $\partial\Omega$ and consider the initial boundary value problem

$$(1.1)\qquad \begin{aligned} &u_t - \Delta u = f, \quad \text{in } \Omega, \quad \text{for } t \geq 0, \\ &u = 0, \quad \text{on } \partial\Omega, \quad \text{for } t \geq 0, \\ &u(x,0) = v(x), \quad \text{in } \Omega, \end{aligned}$$

where u_t denotes $\partial u/\partial t$ and $\Delta = \Sigma_{j=1}^{n}\partial^2/\partial x_j^2$ the Laplacian. In the first step we shall want to approximate $u(x,t)$ by means of a function $u_h(x,t)$ which, for each fixed t, belongs to a finite-dimensional linear space S_h of functions of x with certain approximation properties. This function will be a solution of a finite system of ordinary differential equations and is referred to as a semidiscrete solution of our problem.

Before we turn to the differential equation, we consider briefly the approximation of smooth functions in Ω which vanish on $\partial\Omega$. For concreteness, we shall exemplify by piecewise linear functions in a convex plane domain.

Thus let $\mathcal{T}_h$ denote a partition of Ω into disjoint triangles τ such that no vertex of any triangle lies on the interior of a side of another triangle and such that the union of the triangles determine a polygonal domain $\Omega_h \subset \Omega$ whose boundary vertices lie on $\partial\Omega$.

Let h denote the maximal length of a side of the triangulation $\mathcal{T}_h$. Thus h is a parameter which decreases as the triangulation is made finer. We shall assume that the angles of the triangulations are bounded below, independently of h, and often also that the triangulations are quasi-uniform in the sense that the triangles of $\mathcal{T}_h$ are of essentially the same size, which may be expressed by demanding that the area of τ in $\mathcal{T}_h$ is bounded below by ch^2 with $c>0$ independent of h.

Let now S_h denote the continuous functions on the closure $\bar{\Omega}$ of Ω which are linear in each triangle of $\mathcal{T}_h$ and which vanish outside Ω_h. Let $\{P_j\}_1^{N_h}$ be the interior vertices of $\mathcal{T}_h$. A function in S_h is then uniquely determined by its values at the points P_j and thus depends on N_h parameters. Let Φ_j be the "pyramid function" in S_h which takes the value 1 at P_j

but vanishes at the other vertices. Then $\{\phi_j\}_1^{N_h}$ forms a basis for S_h, and every χ in S_h admits the representation

$$\chi(x) = \sum_{j=1}^{N_h} \alpha_j \phi_j(x), \qquad \text{with } \alpha_j = \chi(P_j).$$

Given a smooth function v on Ω which vanishes on $\partial\Omega$, we can now, for instance, approximate it by its interpolant $I_h v$ in S_h, which we define as the element of S_h which agrees with v at the interior vertices, i.e. $I_h v(P_j) = v(P_j)$ for $j = 1,\ldots,N_h$. We shall need some results concerning the error in this interpolation.

We shall denote below by $\|\cdot\|$ the L_2 or mean square norm over Ω and by $\|\cdot\|_r$ that in the Sobolev space $H^r(\Omega) = W_2^r(\Omega)$. Thus, for real-valued functions v,

$$\|v\| = \left(\int_\Omega v^2 dx\right)^{1/2},$$

and for r a positive integer,

$$\|v\|_r = \left(\sum_{|\alpha|\le r} \|D^\alpha v\|^2\right)^{1/2},$$

where with $\alpha = (\alpha_1,\cdots,\alpha_d)$, $D^\alpha = (\partial/\partial x_1)^{\alpha_1}\cdots(\partial/\partial x_d)^{\alpha_d}$ denotes an arbitrary derivative with respect to x of order $|\alpha| = \sum_{j=1}^{d} \alpha_j$ so that the sum above contains all such derivatives of order at most r. We recall that for functions in $H_0^1(\Omega)$, i.e. the functions v with $\nabla v = \text{grad } v$ in $L_2(\Omega)$ and which vanish on $\partial\Omega$, $\|\nabla v\|$ and $\|v\|_1$ are equivalent norms.

The following error estimates for the interpolant just defined are well-known, namely

$$\|I_h v - v\| \le Ch^2 \|v\|_2,$$

and

$$\|\nabla I_h v - \nabla v\| \le Ch\|v\|_2,$$

where, as will always be the case in the sequel, the statements of the inequalities assume that v is sufficently regular for the norms on the right to be finite.

We shall now return to the general case of a domain Ω in R^d and assume that we are given a family $\{S_h\}$ of finite-dimensional subspaces of $H_0^1(\Omega)$ such that, for some integer $r \ge 2$ and small h,

$$(1.2) \qquad \inf_{\chi\in S_h} \{\|v-\chi\| + h\|\nabla(v-\chi)\|\} \le Ch^s\|v\|_s, \qquad \text{for } 1\le s\le r,$$
$$v \in H^s(\Omega)\cap H_0^1(\Omega).$$

The above example of piecewise linear functions corresponds to $d=r=2$. Also, in the general situation estimates such as (1.2) may often be obtained by exhibiting an interpolation operator I_h into S_h such that

$$(1.3) \qquad \|I_h v-v\| + h\|\nabla(I_h v-v)\| \le Ch^s\|v\|_s, \quad \text{for } 1\le s\le r.$$

For the case that Ω is curved and $r > 2$ there are difficulties near the boundary, but the above situation may be accomplished, in principle, by mapping a curved triangle onto a straight-edged one (isoparametric elements). We shall not dwell on this.

The optimal orders to which functions and their gradients may be approximated under our assumption (1.2) are $O(h^r)$ and $O(h^{r-1})$, respectively, and we shall attempt below to obtain approximations of these orders for the solution of the heat equation.

For the purpose of defining thus an approximate solution to the initial boundary value problem (1.1) we first write this problem in weak form: We multiply the heat equation by a smooth function φ which vanishes on Ω, integrate over Ω, and apply Green's formula to the second term, to obtain, for all such φ, with (v,w) denoting the inner product $\int_\Omega vw dx$ in $L_2(\Omega)$,

$$(u_t,\varphi)+(\nabla u,\nabla\varphi) = (f,\varphi), \qquad \text{for } t \geq 0.$$

We may then pose the approximate problem to find $u_h(t)$, belonging to S_h for each t, such that

$$(1.4) \qquad (u_{h,t},\chi) + (\nabla u_h,\nabla\chi) = (f,\chi), \qquad \forall\, \chi \in S_h, \quad t \geq 0,$$

together with the initial condition

$$u_h(0) = v_h,$$

where v_h is some approximation of v in S_h. Since we have only discretized in the space variables, this is referred to as a semidiscrete problem. Later, we shall discretize also in the time variables to produce completely discrete schemes.

In terms of a basis $\{\Phi_j\}_1^{N_h}$ for S_h our semidiscrete problem may be stated: Find the coefficient $\alpha_j(t)$ in

$$u_h(x,t) = \sum_{j=1}^{N_h} \alpha_j(t)\Phi_j(x),$$

such that

$$\sum_{j=1}^{N_h} \alpha_j'(t)(\Phi_j,\Phi_k) + \sum_{j=1}^{N_h} \alpha_j(t)(\nabla\Phi_j,\nabla\Phi_k) = (f,\Phi_k),$$
$$k = 1,\dots,N_h,$$

and, with γ_j the components of the given initial approximation v_h,

$$\alpha_j(0) = \gamma_j, \qquad j = 1,\dots,N_h.$$

In matrix notation this may be expressed as

$$A\alpha'(t) + B\alpha(t) = \tilde{f}(t), \quad \text{for } t > 0, \text{ with } \alpha(0) = \gamma,$$

where $A = (a_{jk})$ is the mass matrix with elements $a_{jk} = (\phi_j, \phi_k)$, $B = (b_{jk})$ the stiffness matrix with $b_{jk} = (\nabla\phi_j, \nabla\phi_k)$, $\tilde{f} = (f_k)$ the vector with entries $f_k = (f, \phi_k)$, $\alpha(t)$ the vector of unknowns $\alpha_j(t)$ and $\gamma = (\gamma_k)$. The dimension of all these arrays equals N_h, the dimension of S_h.

Since the mass matrix A is a Gram matrix, and thus in particular positive definite and invertible, the above system of ordinary differential equations may be written

$$\alpha'(t) + A^{-1}B\alpha(t) = A^{-1}\tilde{f}, \quad \text{for } t \geq 0, \quad \text{with } \alpha(0) = \gamma,$$

and hence obviously has a unique solution for positive t.

When S_h consists of piecewise polynomial functions, the elements of the matrices A and B may be calculated exactly and once and for all. However, unless f has a particularly simple form, the elements (f, ϕ_j) of $\tilde{f}$ have to be computed by some quadrature formula. We shall consider a special example of this at the end of this section.

We shall prove the following estimate for the error between the solutions of the semidiscrete and continuous problems.

<u>Theorem 1.1</u>. Let u_h and u be the solutions of (1.4) and (1.1), respectively. Then

$$\|u_h(t) - u(t)\| \leq \|v_h - v\| + Ch^r\{\|v\|_r + \int_0^t \|u_t\|_r ds\}, \quad \text{for } t \geq 0.$$

Here we require, of course, that the solution of the continuous problem has the regularity implicitly assumed by the presence of the norms on the right and that v vanishes on Ω. Note also that if (1.3) holds and $v_h = I_h v$, then the first term on the right is dominated by the second. The same holds true if $v_h = P_h v$, where P_h denotes the L_2-projection of v onto S_h, since this choice is the best approximation of v in S_h with respect to the L_2 norm. Another such optimal order choice of v_h is the projection to be defined next.

For the purpose of the proof of Theorem 1.1 we introduce the so called elliptic or Ritz projection R_h onto S_h as the orthogonal projection with respect to the inner product $(\nabla v, \nabla w)$, so that

$$(1.5) \qquad (\nabla R_h v, \nabla \chi) = (\nabla v, \nabla \chi), \qquad \forall\ \chi \in S_h.$$

The function $R_h u$ may also be described as the finite element approximation of the solution of the corresponding elliptic problem, whose exact solution is v. From the well established error analysis for the elliptic problem we quote the following error estimate.

<u>Lemma 1.1</u>. With R_h defined by (1.5) we have

$$\|R_h v - v\| + h\|\nabla(R_h v - v)\| \le Ch^s \|v\|_s, \qquad \text{for } 1 \le s \le r,$$
$$v \in H^s(\Omega) \cap H_0^1(\Omega).$$

Proof. We start with the gradient. We have, using (1.5), for any $\chi \in S_h$,

$$\|\nabla(R_h v-v)\|^2 = (\nabla(R_h v-v),\nabla(R_h v-v)) = (\nabla(R_h v-v),\nabla(\chi-v))$$
$$\leq \|\nabla(R_h v-v)\| \, \|\nabla(\chi-v)\|,$$

and hence, by (1.2),

$$\|\nabla(R_h v-v)\| \leq \inf_{\chi\in S_h} \|\nabla(\chi-v)\| \leq Ch^{s-1}\|v\|_s.$$

For the L_2 norm we proceed by duality. Let φ be arbitrary in $L_2(\Omega)$, take $\psi\in H^2(\Omega)\cap H_0^1(\Omega)$ as the solution of

$$-\Delta\psi = \varphi, \qquad \text{in } \Omega,$$
$$\psi = 0, \qquad \text{on } \partial\Omega,$$

and recall the a priori inequality

$$\|\psi\|_2 \leq C\|\Delta\psi\| = C\|\varphi\|.$$

Then

$$(R_h v-v,\varphi) = -(R_h v-v,\Delta\psi) = (\nabla(R_h v-v),\nabla\psi)$$
$$= (\nabla(R_h v-v),\nabla(\psi-R_h\psi)) \leq \|\nabla(R_h v-v)\|\cdot\|\nabla(\psi-R_h\psi)\|$$
$$\leq Ch^{s-1}\|v\|_s \, h\|\psi\|_2 \leq Ch^s\|v\|_s\|\varphi\|,$$

which completes the proof if we choose $\varphi = R_h v-v$.

We now turn to the proof of Theorem 1.1. In the main step we shall compare the solution of the semidiscrete problem to the elliptic projection of the exact solution. We write

(1.6) $\quad u_h-u = (u_h-R_h u) + (R_h u-u) = \theta + \rho.$

The second term is easily bounded by Lemma 1.1 and obvious estimates:

$$\|\rho(t)\| \leq Ch^r\|u(t)\|_r = Ch^r\|v + \int_0^t u_t ds\|_r$$
$$\leq Ch^r\{\|v\|_r + \int_0^t \|u_t\|_r ds\},$$

In order to estimate θ, we note that

(1.7) $$(\theta_t,\chi) + (\nabla\theta,\nabla\chi)$$
$$= (u_{h,t},\chi) + (\nabla u_h,\nabla\chi) - (R_h u_t,\chi) - (\nabla R_h u,\nabla\chi)$$
$$= (f,\chi) - (R_h u_t,\chi) - (\nabla u,\nabla\chi) = (u_t - R_h u_t,\chi),$$

or

(1.8) $$(\theta_t,\chi) + (\nabla\theta,\nabla\chi) = -(\rho_t,\chi), \qquad \forall\ \chi \in S_h.$$

In this derivation we have used the definition of R_h and the easily established fact that this operator commutes with time differentiation. Since θ belongs to S_h we may choose $\chi = \theta$ in (1.8) and conclude

(1.9) $$(\theta_t,\theta) + \|\nabla\theta\|^2 = -(\rho_t,\theta),$$

or, since the first term equals $\frac{1}{2}\frac{d}{dt}\|\theta\|^2$ and the second is non-negative,

$$\frac{1}{2}\frac{d}{dt}\|\theta\|^2 \le \|\rho_t\|\cdot\|\theta\|.$$

This yields

$$\frac{d}{dt}\|\theta\| \le \|\rho_t\|,$$

or, after integration,

$$\|\theta(t)\| \le \|\theta(0)\| + \int_0^t \|\rho_t\|ds.$$

Here

$$\|\theta(0)\| = \|v_h - R_h v\| \le \|v_h - v\| + \|R_h v - v\|$$
$$\le \|v_h - v\| + Ch^r\|v\|_r,$$

and further

$$\|\rho_t\| = \|R_h u_t - u_t\| \le Ch^r\|u_t\|_r.$$

Together these estimates show the theorem.

In the above proof we made use in (1.9) of the fact that $\|\nabla\theta\|^2$ is non-negative. By a somewhat less wasteful treatment of this term one may demonstrate that the effect of the initial data upon the error tend to zero exponentially as t grows. In fact, with λ_1 the smallest eigenvalue of $-\Delta$ with Dirichlet boundary data, we have

$$\|\nabla\chi\|^2 \geq \lambda_1\|\chi\|^2, \qquad \forall\ \chi \in S_h,$$

and hence (1.9) yields

$$\frac{1}{2}\frac{d}{dt}\|\theta\|^2 + \lambda_1\|\theta\|^2 \leq \|\rho_t\|\cdot\|\theta\|,$$

or

$$\frac{d}{dt}\|\theta\| + \lambda_1\|\theta\| \leq \|\rho_t\|,$$

and

$$\begin{aligned}\|\theta(t)\| &\leq e^{-\lambda_1 t}\|\theta(0)\| + \int_0^t e^{-\lambda_1(t-s)}\|\rho_t(s)\|ds \\ &\leq e^{-\lambda_1 t}\|v_h - v\| + Ch^r\{e^{-\lambda_1 t}\|v\|_r + \int_0^t e^{-\lambda_1(t-s)}\|u_t(s)\|_r ds\}.\end{aligned}$$

Since

$$\|\rho(t)\| \leq Ch^r\|u(t)\|_r,$$

we conclude that

$$\begin{aligned}\|u_h(t)-u(t)\| &\leq e^{-\lambda_1 t}\|v_h - v\| \\ &+ Ch^r\{e^{-\lambda_1 t}\|v\|_r + \|u(t)\|_r + \int_0^t e^{-\lambda_1(t-s)}\|u_t(s)\|_r ds\}.\end{aligned}$$

We shall not pursue this analysis for large t below.

We shall briefly look at another approach to the proof of Theorem 1.1 which consists in working with the equation for θ in operator form. For this purpose we introduce a "discrete Laplacian" Δ_h, which we think of as an operator from S_h into itself, by

$$(\Delta_h\psi,\chi) = -(\nabla\psi,\nabla\chi), \qquad \forall\ \psi,\ \chi \in S_h;$$

this analogue of Green's formula clearly defines $\Delta_h\psi = \sum_{j=1}^{N_h} d_j\phi_j$ from

$$\sum_{j=1}^{N_h} d_j(\phi_j,\phi_k) = -(\nabla\psi,\nabla\phi_k), \quad k = 1,\ldots,N_h,$$

since the matrix of this system is the positive definite mass matrix encountered above. The operator Δ_h is easily seen to be selfadjoint and $-\Delta_h$ is positive definite. Note that Δ_h is related to our other operators by

$$\Delta_h R_h = P_h\Delta. \tag{1.10}$$

For, with $\chi \in S_h$,

$$(\Delta_h R_h v,\chi) = -(\nabla R_h v,\nabla\chi) = -(\nabla v,\nabla\chi) = (P_h\Delta v,\chi).$$

The semidiscrete equation now takes the form

$$(u_{h,t},\chi) - (\Delta_h u_h,\chi) = (P_h f,\chi), \qquad \forall\ \chi \in S_h,$$

or, since the factors on the left are all in S_h,

$$u_{h,t} - \Delta_h u_h = P_h f.$$

Using (1.10) we hence obtain, for θ,

$$\begin{aligned}\theta_t - \Delta_h\theta &= (u_{h,t} - \Delta_h u_h) - (R_h u_t - \Delta_h R_h u)\\ &= P_h f + (P_h - R_h)u_t - P_h(u_t - \Delta u) = P_h(I - R_h)u_t = -P_h\rho_t,\end{aligned}$$

or

$$\theta_t - \Delta_h\theta = -P_h\rho_t. \tag{1.11}$$

Let us denote by $E_h(t)$ the solution operator of the initial value problem for the homogeneous semidiscrete equation,

$$u_{h,t}-\Delta_h u_h = 0, \qquad \text{for } t \geq 0,$$

i.e. the operator which takes the initial data $u_h(0) = v_h$ into the solution $u_h(t)$ at time t, so that $u_h(t) = E_h(t)v_h$. (This operator can also be thought of as the semigroup generated by $-\Delta_h$.) Duhamel's principle then tells us that the solution of the inhomogeneous equation (1.11) is

$$\theta(t) = E_h(t)\theta(0) - \int_0^t E_h(t-s)P_h\rho_t(s)ds.$$

We now note that $E_h(t)$ is stable in L_2, or, more precisely,

$$\|E_h(t)v_h\| \leq \|v_h\|, \qquad \forall\ v_h \in S_h,\ t \geq 0.$$

In fact, choosing $\chi = u_h$ in the homogeneous form of (1.4) we have

$$\frac{1}{2}\frac{d}{dt}\|u_h\|^2 + \|\nabla u_h\|^2 = 0.$$

Since the second term is non-negative we conclude that the first is non-positive and hence $\|u_h\|^2$ non-increasing, which shows the stability. Since obviously P_h has unit norm in L_2 we have therefore

$$\|\theta(t)\| \leq \|\theta(0)\| + \int_0^t \|\rho_t(s)\|ds,$$

from which the theorem follows as above. The desired estimate for θ is thus a consequence of the stability estimate for $E_h(t)$ combined with the error estimate for the elliptic problem applied to $\rho_t = (R_h-I)u_t$.

In a similar way we may prove the following estimate for the error in the gradient.

Theorem 1.2. Under the assumptions of Theorem 1.1 we have, for $t \geq 0$,

$$\|\nabla u_h(t)-\nabla u(t)\|$$
$$\leq C\|\nabla v_h - \nabla v\| + Ch^{r-1}\{\|v\|_r + \|u(t)\|_r + (\int_0^t \|u_t\|_{r-1}^2 ds)^{1/2}\}.$$

Proof. As before we write the error in the form (1.6). Here, by Lemma 1.1,

$$\|\nabla\rho(t)\| = \|\nabla(R_h u(t)-u(t))\| \leq Ch^{r-1}\|u(t)\|_r.$$

In order to estimate $\nabla\theta$ we use again (1.8), now with $\chi = \theta_t$. We obtain

$$\|\theta_t\|^2 + \frac{1}{2}\frac{d}{dt}\|\nabla\theta\|^2 = -(\rho_t,\theta_t) \leq \frac{1}{2}\|\rho_t\|^2 + \frac{1}{2}\|\theta_t\|^2,$$

so that

$$\frac{d}{dt}\|\nabla\theta\|^2 \leq \|\rho_t\|^2$$

or

$$\|\nabla\theta(t)\|^2 \leq \|\nabla\theta(0)\|^2 + \int_0^t \|\rho_t\|^2 ds$$
$$\leq (\|\nabla(v_h-v)\| + \|\nabla(R_h v-v)\|)^2 + \int_0^t \|\rho_t\|^2 ds$$

Hence, in view of Lemma 1.1,

$$\|\nabla\theta(t)\|^2 \leq C\|\nabla(v_h-v)\|^2 + Ch^{2r-2}\{\|v\|_r^2 + \int_0^t \|u_t\|_{r-1}^2 ds\}, \tag{1.12}$$

which completes the proof.

Note that if $v_h = I_h v$ as in (1.3), or $v_h = P_h v$, then

$$\|\nabla(v_h - v)\| \le Ch^{r-1}\|v\|_r,$$

so that the first term on the right in Theorem 1.2 is again dominated by the second.

In the case that the triangulation $\mathcal{T}_h$ is quasiuniform, an estimate for the error in the gradient may also be obtained directly from the result of Theorem 1.1 by means of the "inverse" inequality

$$(1.13) \qquad \|\nabla\chi\| \le Ch^{-1}\|\chi\|, \qquad \forall\ \chi \in S_h.$$

This inequality follows easily from the corresponding inequality for each triangle $\tau \in \mathcal{T}_h$, which in turn is obtained by a transformation to a fixed reference triangle, and using the fact that all norms on a finite dimensional space are equivalent, see e.g. Ciarlet [7]. Using (1.13) we have, for χ arbitrary in S_h,

$$\begin{aligned}\|\nabla u_h(t) - \nabla u(t)\| &\le \|\nabla u_h(t) - \nabla\chi\| + \|\nabla\chi - \nabla u(t)\| \\ &\le Ch^{-1}\{\|u_h(t) - \chi\| + h\|\nabla\chi - \nabla u(t)\|\} \\ &\le Ch^{-1}\|u_h(t) - u(t)\| + Ch^{-1}\{\|\chi - u(t)\| + h\|\nabla\chi - \nabla u(t)\|\}.\end{aligned}$$

Here, by our approximation assumption (1.2), we have, with suitable choice of χ,

$$\|\chi - u(t)\| + h\|\nabla\chi - \nabla u(t)\| \le Ch^r\|u(t)\|_r,$$

and hence, by Theorem 1.1, for the appropriate choice of v_h,

$$\|\nabla u_h(t) - \nabla u(t)\| \le Ch^{r-1}(\|v\|_r + \int_0^t \|u_t(s)\|_r ds).$$

We make the following observation concerning the estimate for the gradient of $\theta = u_h - R_h u$ in the proof of Theorem 1.2: Assume that we have chosen $v_h = R_h v$ so that $\theta(0) = 0$. Then, in addition to (1.12), we have

$$(1.14) \qquad \|\nabla\theta(t)\| \le C\left(\int_0^t \|\rho_t\|^2 ds\right)^{1/2} \le Ch^r\left(\int_0^t \|u_t\|_r^2 ds\right)^{1/2}.$$

Hence the gradient of θ is $O(h^r)$, whereas the gradient of the total error is only $O(h^{r-1})$, for small h. Thus ∇u_h is a better approximation to $\nabla R_h u$ than is possible to ∇u. This is an example of a phenomenon which is sometimes referred to as superconvergence (cf. Křížek and Neittanmäki [20] and references).

As a simple example of an application of such an estimate we shall indicate briefly how it may be used to show an essentially optimal order error bound in the maximum-norm. Consider thus the concrete situation described in the beginning of this chapter with Ω a plane smooth convex domain and S_h consisting of piecewise linear functions ($d = r = 2$) on a quasi-uniform triangulation of Ω. It is then the case, as was proved in Nitsche [26], that the error in the elliptic problem may be bounded in such a way that

$$\|\rho(t)\|_{L_\infty(\Omega)} = \|R_h u(t) - u(t)\|_{L_\infty(\Omega)} \le Ch^2 \log\frac{1}{h} \|u(t)\|_{W^2_\infty(\Omega)}.$$

In two dimensions, Sobolev's inequality almost bounds the maximum-norm by the norm in $H^1(\Omega)$, and it may be shown, using an inverse assumption of a type similar to (1.13), that, for functions in the subspace S_h,

$$\|\chi\|_{L_\infty(\Omega)} \le C\left(\log\frac{1}{h}\right)^{1/2} \|\nabla\chi\|.$$

Applied to θ this shows, by (1.14) (with r=2), that

$$\|\theta(t)\|_{L_\infty(\Omega)} \le C(\log\frac{1}{h})^{1/2}\, h^2 (\int_0^t \|u_t\|_2^2 ds)^{1/2},$$

and we may thus conclude for the error in the parabolic problem that

$$\|u_h(t)-u(t)\|_{L_\infty(\Omega)} \le \|\rho(t)\|_{L_\infty(\Omega)} + \|\theta(t)\|_{L_\infty(\Omega)}$$
$$\le C(u,t)h^2\log\frac{1}{h}.$$

We shall now turn our attention to some simple schemes for discretization also with respect to the time variable. We begin with the backward Euler-Galerkin method. Letting k be the time step and U^n the approximation in S_h of u(t) at $t = t_n = nk$, this method is defined by replacing the time derivative in (1.4) by a backward difference quotient, or, setting $\bar{\partial}_t U^n = k^{-1}(U^n-U^{n-1})$,

(1.15) $$(\bar{\partial}_t U^n,\chi) + (\nabla U^n,\nabla\chi) = (f(t_n),\chi), \quad \forall\ \chi \in S_h, \quad n \ge 1,$$
$$U^0 = v_h.$$

This defines U^n implicitly by means of U^{n-1} from the elliptic problem

$$(U^n,\chi) + k(\nabla U^n,\nabla\chi) = (U^{n-1} + kf(t_n),\chi), \quad \forall\ \chi \in S_h.$$

Note that, with notation analogous to the semidiscrete situation, this may be written

$$(A + kB)\alpha^n = A\alpha^{n-1} + k\tilde{f}(t_n),$$

where A+kB is positive definite and hence, in particular, invertible.

We shall prove the following error estimate:

<u>Theorem 1.3</u>. With U^n and u the solutions of (1.15) and (1.1), respectively, we have

$$\|U^n - u(t_n)\| \le \|v_h - v\| + Ch^r\{\|v\|_r + \int_0^{t_n}\|u_t\|_r ds\} + k\int_0^{t_n}\|u_{tt}\|ds, \quad \text{for } n \ge 0.$$

<u>Proof</u>. In analogy with (1.6) we write

$$U^n-u(t_n) = (U^n-R_hu(t_n)) + (R_hu(t_n)-u(t_n)) = \theta^n + \rho^n.$$

As before,

$$\|\rho^n\| \le Ch^r\|u(t_n)\|_r \le Ch^r\{\|v\|_r + \int_0^{t_n}\|u_t\|_r ds\}. \tag{1.16}$$

This time, a calculation corresponding to (1.7) yields

$$(\bar{\partial}_t\theta^n,\chi) + (\nabla\theta^n,\nabla\chi) = -(\omega^n,\chi),$$

where

$$\omega^n = R_h\bar{\partial}_t u(t_n) - u_t(t_n)$$
$$= (R_h-I)\bar{\partial}_t u(t_n) + (\bar{\partial}_t u(t_n) - u_t(t_n)) = \omega_1^n + \omega_2^n.$$

Choosing $\chi = \theta^n$ we have

$$(\bar{\partial}_t\theta^n,\theta^n) \le \|\omega^n\|\cdot\|\theta^n\|,$$

or

$$\|\theta^n\|^2 - (\theta^{n-1},\theta^n) \le k\|\omega^n\|\cdot\|\theta^n\|,$$

so that

$$\|\theta^n\| \le \|\theta^{n-1}\| + k\|\omega^n\|,$$

and, by repeated application,

$$\|\theta^n\| \le \|\theta^0\| + k\sum_{j=1}^n \|\omega^j\| \le \|\theta^0\| + k\sum_{j=1}^n \|\omega_1^j\| + k\sum_{j=1}^n \|\omega_2^j\|.$$

Here, as before,

$$\|\theta^0\| = \|v_h-R_hv\| \le \|v_h-v\| + Ch^r\|v\|_r.$$

Note now that

$$\omega_1^j = (R_h - I)k^{-1}\int_{t_{j-1}}^{t_j} u_t\,ds = k^{-1}\int_{t_{j-1}}^{t_j} (R_h - I)u_t\,ds,$$

whence

$$k\sum_{j=1}^{n} \|\omega_1^j\| \le \sum_{j=1}^{n}\int_{t_{j-1}}^{t_j} Ch^r\|u_t\|_r\,ds = Ch^r\int_0^{t_n} \|u_t\|_r\,ds.$$

Further,

$$\omega_2^j = k^{-1}(u(t_j)-u(t_{j-1})) - u_t(t_j)$$
$$= -k^{-1}\int_{t_{j-1}}^{t_j} (s-t_{j-1})u_{tt}(s)\,ds,$$

so that

$$k\sum_{j=1}^{n} \|\omega_2^j\| \le \sum_{j=1}^{n} \left\|\int_{t_{j-1}}^{t_j} (s-t_{j-1})u_{tt}(s)\,ds\right\| \le k\int_0^{t_n} \|u_{tt}\|\,ds.$$

Together our estimates complete the proof of the theorem.

Notice that because of the non-symmetric choice of the discretization in time, the backward Euler-Galerkin method is only first order accurate in time.

We now turn to the Crank-Nicolson-Galerkin method, in which the semidiscrete equation is discretized in a symmetric fashion around the point $t_{n-1/2} = (n-1/2)k$, and which will therefore produce a second order in time accurate method. More precisely, we define U^n in S_h recursively by

$$(1.17)\qquad (\bar{\partial}_t U^n,\chi) + (\nabla(U^n+U^{n-1})/2,\nabla\chi) = (f(t_{n-1/2}),\chi), \quad \forall\ \chi \in S_h,\ n \ge 1,$$

$$U^0 = v_h.$$

This time the equation of U^n may be written in matrix form as

$$(A + \tfrac{1}{2}kB)\alpha^n = (A - \tfrac{1}{2}kB)\alpha^{n-1} + k\tilde{f}(t_{n-1/2}),$$

with the matrix $A + \frac{1}{2}kB$ positive definite.

Now the error estimate reads as follows.

Theorem 1.4. With U^n and u the solutions of (1.17) and (1.1), respectively, we have

$$\|U^n - u(t_n)\| \le \|v_h - v\| + Ch^r\{\|v\|_r + \int_0^{t_n} \|u_t\|_r ds\}$$
$$+ k^2 \int_0^{t_n} (\|u_{ttt}\| + \|\Delta u_{tt}\|) ds, \quad \text{for } n \ge 0.$$

Proof. We first recall that (1.16) holds so that it remains only to estimate θ^n. With the above notation we have

$$(\bar{\partial}_t \theta^n, \chi) + (\nabla(\theta^n + \theta^{n-1})/2, \nabla\chi) = -(\omega^n, \chi),$$

where

$$\omega^n = (R_h - I)\bar{\partial}_t u(t_n) + (\bar{\partial}_t u(t_n) - u_t(t_{n-1/2}))$$
$$+ \Delta(u(t_{n-1/2}) - \tfrac{1}{2}(u(t_n) + u(t_{n-1}))) = \omega_1^n + \omega_2^n + \omega_3^n.$$

Choosing this time $\chi = (\theta^n + \theta^{n-1})/2$ in the equation for θ, we find

$$(\bar{\partial}_t \theta^n, (\theta^n + \theta^{n-1})/2) \le \tfrac{1}{2}\|\omega^n\|(\|\theta^n\| + \|\theta^{n-1}\|),$$

or

$$\|\theta^n\|^2 - \|\theta^{n-1}\|^2 \le k\|\omega^n\|(\|\theta^n\| + \|\theta^{n-1}\|),$$

or, after cancellation of a common factor,

$$\|\theta^n\| \le \|\theta^{n-1}\| + k\|\omega^n\|.$$

By repeated application this yields

$$\|\theta^n\| \le \|\theta^0\| + k\sum_{j=1}^{n} (\|\omega_1^j\| + \|\omega_2^j\| + \|\omega_3^j\|).$$

Since θ^0 may be estimated as above, it remains to bound the latter sum. Here, as before,

$$\|\omega_1^j\| = \|(R_h-I)\bar{\partial}_t u(t_j)\| \le Ch^r k^{-1}\int_{t_{j-1}}^{t_j} \|u_t\|_r ds.$$

Further,

$$\|\omega_2^j\| = \|\bar{\partial}_t u(t_j) - u_t(t_{j-1/2})\|$$

$$= \tfrac{1}{2}k^{-1}\|\int_{t_{j-1}}^{t_{j-1/2}} (s-t_{j-1})^2 u_{ttt}(s)ds + \int_{t_{j-1/2}}^{t_j} (s-t_j)^2 u_{ttt}(s)ds\|$$

$$\le Ck\int_{t_{j-1}}^{t_j} \|u_{ttt}\|ds,$$

and similarly

$$\|\omega_3^j\| = \|\Delta(u(t_{j-1/2}) - \tfrac{1}{2}(u(t_j) + u(t_{j-1})))\|$$

$$\le Ck\int_{t_{j-1}}^{t_j} \|\Delta u_{tt}\|ds.$$

Altogether,

$$k\sum_{j=1}^{n} (\|\omega_1^j\| + \|\omega_2^j\| + \|\omega_3^j\|)$$

$$\le Ch^r\int_0^{t_n} \|u_t\|_r ds\} + k^2\int_0^{t_n} (\|u_{ttt}\|+\|\Delta u_{tt}\|)ds,$$

which completes the proof.

The above material, extracted here from Thomée [40], is now standard; we refer to Douglas and Dupont [12], Price and Varga [30], and Fix and Nassif [14] for some early references. The use of the Ritz projection of the exact solution as a tool in the error analysis was initiated in Wheeler [44]. For generalization to more general parabolic operators, in a framework that permits finite element spaces where the functions do not necessarily vanish on the boundary, see Thomée [40] and references quoted in this work. For error analyses in maximum-norm, see also Dobrowolski [10], Nitsche [27], Nitsche and Wheeler [28], Schatz, Thomée, and Wahlbin [37], Thomée and Wahlbin [42], and Wheeler [45].

We return briefly to the computation of the quantities (f,χ) on the right hand side of (1.4). We shall consider only the case of piecewise linear approximating functions in the plane, described at the outset. For this we shall use a quadrature formula, on each triangle τ of the triangulation.

Let thus $P_{\tau,j}$, $j=1,2,3$, be the vertices of τ and set

$$Q_{\tau,h}(g) = \frac{1}{3}\,\text{area}(\tau)\sum_{j=1}^{3} g(P_{\tau,j}) \simeq \int_\tau g\,dx. \tag{1.18}$$

We may then define an approximation of the inner product in $L_2(\Omega)$ by

$$(\psi,\chi)_h = \sum_{\tau\in\mathcal{T}_h} Q_{\tau,h}(\psi\chi).$$

With this the natural choice for the approximation of (1.4) is

$$(u_{h,t},\chi) + (\nabla u_h,\nabla\chi) = (f,\chi)_h, \quad \forall\ \chi \in S_h, \text{ for } t \geq 0. \tag{1.19}$$

Note that f now enters only at the vertices of the triangulation. It is not difficult to see that this method is of optimal order $O(h^2)$, under the appropriate regularity assumptions.

Instead of analyzing the method (1.19) here we shall carry the use of the quadrature one step further and consider the method defined by

$$(u_{h,t},\chi)_h + (\nabla u_h,\nabla\chi) = (f,\chi)_h, \quad \forall\ \chi \in S_h, \text{ for } t \geq 0, \tag{1.20}$$

thus employing quadrature also in the first term. As we have pointed out above, this is not needed from the point of view of calculating the coefficients of the mass matrix A associated with (1.4), but has other advantages, which we shall now briefly comment upon. Firstly, writing (1.20) in matrix form we have now

$$\bar{A}\alpha' + B\alpha = \bar{f}, \quad \text{for} \quad t \geq 0,$$

where B is as before, but where the elements $\bar{a}_{ij}$ of $\bar{A}$ are defined by

$$\bar{a}_{ij} = (\phi_i,\phi_j)_h.$$

We find at once that $\bar{a}_{ij} = 0$ for $i \neq j$, so that $\bar{A}$ is a positive diagonal matrix, and a simple calculation (see [40], pp. 167-168) shows that

$$\bar{a}_{ii} = \sum_{j=1}^{N_h} a_{ij}.$$

Thus $\bar{A}$ is the diagonal matrix whose diagonal elements are the row sums of the original mass matrix A. The method is therefore referred to as the lumped mass method.

Since $\bar{A}$ is diagonal the associated forward Euler lumped mass method,

$$\left(\frac{U^{n+1}-U^n}{k},\chi\right)_h + (\nabla U^n,\nabla\chi) = (f^n,\chi)_h, \quad \forall\ \chi \in S_h, \quad \text{for } n \geq 0,$$

may be written

$$\bar{A}U^{n+1} = (\bar{A} - kB)U^n + k\bar{f}^n, \qquad \text{for } n \geq 0.$$

As $\bar{A}$ is trivial to invert, this method has the character of an explicit finite difference scheme. This is in contrast to the forward Euler method associated with the standard Galerkin method, where the more difficult to invert matrix A appears instead of $\bar{A}$ on the left hand side.

One may also show that, under an extra assumption, the lumped mass method satisfies a maximum-principle analogous to the one valid for the continuous problem and also for the certain standard finite difference schemes: If all angles of $\mathcal{T}_h$ are non-obtuse ($\le \pi/2$), and if $f = 0$, then for the solution $u_h(x,t) = \bar{E}_h(t)v_h$ hold the inequalities

$$\min(0,\min_{\Omega} v_h(x)) \le u_h(x,t) \le \max(0,\max_{\Omega} v_h(x)), \quad \text{for } t \ge 0. \tag{1.21}$$

In particular, $\bar{E}_h(t)$ is stable in the maximum-norm, and

$$\|\bar{E}_h(t)v_h\|_{L_\infty} \le \|v_h\|_{L_\infty}.$$

The corresponding conclusion holds, e.g., for the backward Euler discretization of (1.8).

One may show the following error estimate, which, because of the use of quadrature, requires a little more regularity than our previous error estimate of Theorem 1.1.

<u>Theorem 1.5</u>. Under the appropriate regularity assumptions we have for the error in (1.20)

$$\|u_h(t)-u(t)\| \le C\|v_h-v\| + Ch^2\{\|v\|_2 + \|u(t)\|_2 + (\int_0^t (\|u_t\|_2^2 + \|f\|_2^2)ds)^{1/2}\}.$$

For the proof we shall need some estimates for the quadrature error

$$\epsilon_h(v,w) = (v,w)_h - (v,w).$$

Lemma 1.2. We have

(1.22) $$|\epsilon_h(\psi,\chi)| \le Ch^2\|\nabla\psi\|\cdot\|\nabla\chi\|, \qquad \text{for } \psi,\ \chi \in S_h$$

and

(1.23) $$|\epsilon_h(\varphi,\chi)| \le Ch^2\|\varphi\|_2\cdot\|\nabla\chi\|, \qquad \text{for } \varphi \in H^2(\Omega),\ \chi \in S_h.$$

Proof. Since the quadrature formula (1.18) is easily seen to be exact for g linear, we have, by transformation to a fixed reference triangle τ_0 and using the Bramble-Hilbert lemma and the Sobolev inequality

$$\|g\|_{L_\infty(\tau_0)} \le C\|g\|_{W_1^2(\tau_0)},$$

that

$$|Q_{\tau,h}(g) - \int_\tau g\,dx| \le Ch^2 \sum_{|\alpha|=2} \|D^\alpha g\|_{L_1(\tau)}.$$

In particular, if ψ and χ are linear in τ,

$$|Q_{\tau,h}(\psi\chi) - \int_\tau \psi\chi\,dx| \le Ch^2\|\nabla\psi\|_{L_2(\tau)}\cdot\|\nabla\chi\|_{L_2(\tau)},$$

from which (1.22) follows by summation and the Cauchy-Schwarz inequality. Similarly, for $\varphi \in H^2(\Omega)$ and χ linear in τ,

$$|Q_{\tau,h}(\varphi\chi) - \int_\tau \varphi\chi\,dx|$$
$$\le Ch^2\{\sum_{|\alpha|=2} \|D^\alpha\varphi\|_{L_2(\tau)}\cdot\|\chi\|_{L_2(\tau)} + \|\nabla\varphi\|_{L_2(\tau)}\cdot\|\nabla\chi\|_{L_2(\tau)}\}$$
$$\le Ch^2\|\varphi\|_{H^2(\tau)}\cdot\|\chi\|_{H^1(\tau)},$$

which implies (1.23).

We can now give the proof of Theorem 1.5. With our usual splitting of the error we have

(1.24) $$\|\rho(t)\| = \|(R_h - I)u(t)\| \le Ch^2\|u(t)\|_2.$$

For θ we have

$$(\theta_t,\chi)_h + (\nabla\theta,\nabla\chi) = -(\rho_t,\chi) + \epsilon_h(f,\chi) - \epsilon_h(R_h u_t,\chi).$$

Choosing $\chi = \theta$ and applying Lemma 1.2, this yields, with $\|\cdot\|_h = (\cdot,\cdot)_h^{1/2}$,

$$\frac{1}{2}\frac{d}{dt}\|\theta\|_h^2 + \|\nabla\theta\|^2 \le \|\rho_t\|\cdot\|\theta\| + Ch^2(\|f\|_2 + \|\nabla R_h u_t\|)\|\nabla\theta\|.$$

Applying (1.24) to u_t and noting that R_h is bounded with respect to $\|\nabla\cdot\|$, we conclude that

$$\text{(1.25)} \qquad \frac{d}{dt}\|\theta\|_h^2 \le Ch^4(\|u_t\|_2^2 + \|f\|_2^2).$$

By transformation to a reference triangle we find easily that $\|\cdot\|_{L_2(\tau)}$ and $Q_{\tau,h}(\cdot,\cdot)^{1/2}$ are equivalent for χ linear, and hence that $\|\cdot\|$ and $\|\cdot\|_h$ are equivalent norms on S_h, uniformly in h. Hence we obtain by integration of (1.25)

$$\|\theta(t)\|^2 \le C\|\theta(0)\|^2 + Ch^4\int_0^t (\|u_t\|_{H^2}^2 + \|f\|_{H^2}^2)\,ds,$$

from which the proof is easily completed.

The present discussion of quadrature and lumping is adopted from Chen and Thomée [6] (cf. also [40]). The maximum principle of (1.22) is from Fujii [15]. For more general results concerning quadrature in the context of finite element methods, see Raviart [34].

2. NON-SMOOTH DATA ERROR ESTIMATES FOR THE SEMIDISCRETE PROBLEM

Recall that the main error estimate for the semidiscrete problem (1.4), Theorem 1.1, shows that under the appropriate regularity assumptions on the exact solution of (1.1), and with a suitable choice of v_h, we have

$$\|u_h(t) - u(t)\| \le Ch^r\{\|v\|_r + \int_0^t \|u_t\|_r ds\}, \quad \text{for } t \ge 0.$$

We shall now discuss the regularity requirement for optimal order convergence in a little more detail, in the case of the homogeneous equation, i.e., for the initial boundary value problem

$$\begin{aligned} &u_t = \Delta u, && \text{in } \Omega, && \text{for } t > 0, \\ &u = 0, && \text{on } \partial\Omega, && \text{for } t > 0, \\ &u(\cdot,0) = v, && \text{in } \Omega. && \end{aligned} \tag{2.1}$$

For this purpose we shall introduce certain function spaces as follows: Let $\{\lambda_m\}_1^\infty$ and $\{\varphi_m\}_1^\infty$ be the eigenvalues and corresponding eigenfunctions of the positive definite operator on $L_2(\Omega)$ defined by $-\Delta$, with Dirichlet boundary conditions. The λ_m are positive and arranged in increasing order, and the φ_m are assumed to be normalized so that they form an orthonormal basis for $L_2(\Omega)$. In particular, any $v \in L_2(\Omega)$ may be represented as

$$v = \sum_{m=1}^{\infty} (v,\varphi_m)\varphi_m,$$

and Parseval's relation,

$$(v,w) = \sum_{m=1}^{\infty} (v,\varphi_m)(w,\varphi_m),$$

holds.

For $s \geq 0$, let $\dot{H}^s = \dot{H}^s(\Omega)$ be the subspace of $L_2(\Omega)$ defined by

$$\|v\|_{\dot{H}^s} = \Big(\sum_{m=1}^{\infty} \lambda_m^s (v,\varphi_m)^2\Big)^{1/2} < \infty.$$

One has the following well-known characterization.

Lemma 2.1. For s a non-negative integer,

$$\dot{H}^s(\Omega) = \{v \in H^s(\Omega);\ \Delta^j v = 0 \text{ on } \partial\Omega \text{ for } j<s/2\}.$$

Proof. See e.g. Thomée [40], p. 34-35.

The solution of the initial-boundary value problem (2.1) may now be represented, with $E(t)$ the associated solution operator, as

$$u(\cdot,t) = E(t)v = \sum_{m=1}^{\infty} e^{-t\lambda_m}(v,\varphi_m)\varphi_m,$$

and one has the following regularity result.

Lemma 2.2. For $v \in L_2(\Omega)$ and $t > 0$, the solution $E(t)v$ of (2.1) belongs to $\dot{H}^s(\Omega)$ for any $s \geq 0$. If $0 \leq \ell \leq s$ and $v \in \dot{H}^\ell(\Omega)$, we have

$$\|E(t)v\|_{\dot{H}^s} \leq Ct^{-(s-\ell)/2}\|v\|_{\dot{H}^\ell}, \qquad \text{for } t > 0.$$

Proof. We have

$$\|u(t)\|^2_{\dot{H}^s} = \sum_{m=1}^{\infty} \lambda_m^s e^{-2\lambda_m t}(v,\varphi_m)^2$$

$$= t^{-(s-\ell)} \sum_{m=1}^{\infty} \lambda_m^{\ell}(t\lambda_m)^{s-\ell} e^{-2\lambda_m t} (v,\varphi_m)^2$$

$$\le Ct^{-(s-1)}\|v\|^2_{\dot{H}^{\ell}},$$

since $y^{s-\ell}e^{2y} \le C$ for $y \ge 0$.

We now return to the discussion of the semidiscrete approximation of our initial-value problem (2.1), and assume that this is based on a family of finite dimensional spaces S_h such that (1.2) holds. We shall demonstrate the following result.

Theorem 2.1. Assume that (1.2) holds, that $v \in \dot{H}^r$, and that v_h is chosen such that

$$\|v_h - v\| \le Ch^r\|v\|_{\dot{H}^r}.$$

Then

$$\|u_h(t) - u(t)\| \le Ch^r\|v\|_{\dot{H}^r}, \quad \text{for } t \ge 0.$$

Proof. We set $T = (-\Delta)^{-1}$, where $-\Delta$ is the positive definite operator introduced above, and let T_h: $L_2(\Omega) \to S_h$ be its discrete analogue, defined by

$$(\nabla T_h f,\nabla\chi) = (f,\chi), \quad \forall\ \chi \in S_h.$$

The operator T_h thus associates with f the finite element solution $T_h f \in S_h$ of the elliptic problem

$$-\Delta u = f, \quad \text{in } \Omega,$$

$$u = 0, \quad \text{on } \partial\Omega.$$

As a result, $T_h(-\Delta)u = R_h u$, where R_h is the Ritz projection introduced in Section 1. The operator T_h is self-adjoint in $L_2(\Omega)$ because, by our definitions,

(2.2) $$(f, T_h g) = (\nabla T_h f, \nabla T_h g), \quad \forall\ f, g \in L_2(\Omega),$$

and since the right hand side is symmetric in f and g, so is the left. Further, it is positive semidefinite because for f = g we obtain

(2.2) $$(f, T_h f) = \|\nabla T_h f\|^2 \geq 0.$$

Finally, restricted to S_h, considered as a subspace of $L_2(\Omega)$, it is positive definite. For, if $\chi \in S_h$ and $(\chi, T_h\chi) = 0$, then $\nabla T_h\chi = 0$ by (2.2). Hence $T_h\chi = 0$, since $T_h\chi$ vanishes on $\partial\Omega$, and thus

$$\|\chi\|^2 = (\nabla T_h\chi, \nabla\chi) = 0,$$

which proves our claim.

In terms of the operator T_h, our discrete equation

$$(u_{h,t}, \chi) + (\nabla u_h, \nabla\chi) = 0,$$

may be written

$$T_h u_{h,t} + u_h = 0, \quad \text{for } t \geq 0,$$

and, similarly, our original equation may be put into the form

$$Tu_t + u = 0, \quad \text{for } t \geq 0.$$

Introducing the error $e = u_h - u$, we have therefore

$$T_h e_t + e = T_h u_{h,t} + u_h - T_h u_t - u$$
$$= -(T_h - T)u_t = -(T_h - T)\Delta u = -(I - R_h)u,$$

or

$$(2.3)\qquad T_h e_t + e = \rho \equiv -(I - R_h)u.$$

Recall that our estimate of Lemma 1.1 for the error in the elliptic projection R_h shows that

$$(2.4)\qquad \|\rho\| = \|(I-R_h)u\| \le Ch^s\|u\|_{\dot{H}^s}, \quad \text{for } 0 \le s \le r.$$

The proof of the error bound of Theorem 2.1 will now be based on the following lemma.

Lemma 2.3. We have

$$\|e(t)\|^2 \le C\{\|\rho(t)\|^2 + t^{-1}\int_0^t (\|\rho\|^2 + s^2\|\rho_t\|^2)\,ds\}.$$

Assuming this for a moment, we may easily complete the proof. In fact, by (2.4) and Lemma 2.2,

$$\|\rho(t)\| \le Ch^r\|u(t)\|_{\dot{H}^r} \le Ch^r\|v\|_{\dot{H}^r}.$$

Further, using Lemma 2.2.

$$t^{-1}\int_0^t \|\rho\|^2 ds \le Ch^{2r}t^{-1}\int_0^t \|u(s)\|^2_{\dot{H}^r} ds$$
$$\le Ch^{2r}t^{-1}\int_0^t \|v\|^2_{\dot{H}^r} ds = Ch^{2r}\|v\|^2_{\dot{H}^r},$$

and similarly, since, again by (2.4) and Lemma 2.2,

$$\|\rho_t\| = \|(I-R_h)u_t\| \le Ch^r\|u_t\|_{\dot{H}^r} \le Ch^r\|\Delta u\|_{\dot{H}^r} \le Ch^r\|u\|_{\dot{H}^{r+2}},$$

we find, using Lemma 2.2 once more,

$$t^{-1}\int_0^t s^2\|\rho_t\|^2 ds \le Ch^{2r}t^{-1}\int_0^t s^2\|u_t(s)\|^2_{\dot{H}^r} ds$$
$$\le Ch^{2r}t^{-1}\int_0^t \|v\|^2_{\dot{H}^r} ds \le Ch^{2r}\|v\|^2_{\dot{H}^r}.$$

Together, these estimates complete the proof.

It remains to show Lemma 2.3. By multiplication of (2.3) by e_t we find

$$(T_h e_t, e_t) + \frac{1}{2}\frac{d}{dt}\|e\|^2 = -(\rho, e_t) = -\frac{d}{dt}(\rho, e) + (\rho_t, e).$$

Hence we derive, using the fact that T_h is positive semidefinite,

$$\frac{d}{dt}(t\|e\|^2) \le -2\frac{d}{dt}(t(\rho,e)) + 2t(\rho_t, e) + \|e\|^2 + 2(\rho, e),$$

whence, by integration over $(0,t)$,

$$t\|e\|^2 \le 2t\|\rho\|\ \|e\| + \int_0^t [\|e\|^2 + 2\|\rho\|\cdot\|e\| + 2s\|\rho_t\|\cdot\|e\|]ds$$

and

$$\|e\|^2 \le C(\|\rho(t)\|^2 + t^{-1}\int_0^t [\|e\|^2 + \|\rho\|^2 + s^2\|\rho_t\|^2]ds.$$

We shall show that

$$\int_0^t \|e\|^2 ds \le \int_0^t \|\rho\|^2 ds, \tag{2.5}$$

which then makes the proof complete. For this, we first note that $T_h e(0) = 0$. In fact, $e(0) = P_h v - v$ and hence, for any $\varphi \in L_2(\Omega)$,

$$(T_h e(0), \varphi) = (P_h v - v, T_h \varphi) = 0,$$

since $T_h\varphi \in S_h$. Now, multiplying (2.3), this time by e, we find

$$(T_h e_t, e) + \|e\|^2 = -(\rho, e) \le \frac{1}{2}\|\rho\|^2 + \frac{1}{2}\|e\|^2.$$

By the symmetry of T_h, the first term may be written as $\frac{1}{2}\frac{d}{dt}(T_h e, e)$, and we obtain hence

$$\frac{d}{dt}(T_h e, e) + \|e\|^2 \le \|\rho\|^2.$$

Since $T_h e(0) = 0$, this implies (2.5) by integration over $(0,t)$.

We now turn to the situation when v is not regular enough to belong to $\dot{H}^r$. Recall from Lemma 2.1 that for v to do so requires, in addition to smoothness, also certain boundary conditions. We shall show that in this case, if v_h is chosen as the L_2-projection $P_h v$ of v onto S_h, we still have optimal order convergence for t bounded away from zero. We emphasize that this time we make a more specific choice of v_h than before.

Theorem 2.2. Assume that (1.2) holds, that $v \in L_2(\Omega)$, and that $v_h = P_h v$. Then

$$\|u_h(t) - u(t)\| \leq Ch^r t^{-r/2}\|v\|, \quad \text{for} \quad t > 0.$$

Proof. Set $u_h(t) = E_h(t)v_h = E_h(t)P_h v$ and define the error operator $F_h(t)$ by

$$e(t) = F_h(t)v = E_h(t)P_h v - E(t)v.$$

We want to prove that, in operator norm on $L_2(\Omega)$,

$$\|F_h(t)\| \leq Ch^r t^{-r/2}. \tag{2.6}$$

In the first step we show this inequality with r replaced by 1. In fact, in this case (2.6) follows easily by Lemmas 2.3 and 2.2. For

$$\|\rho(t)\| \leq Ch\|u(t)\|_{\dot{H}^1} \leq Cht^{-1/2}\|v\|,$$

and

$$\int_0^t \|\rho\|^2 ds \leq Ch^2 \int_0^t \|u(s)\|^2_{\dot{H}^1} ds$$
$$= Ch^2 \int_0^t \sum_{m=1}^{\infty} \lambda_m e^{-2\lambda_m s}(v,\varphi_m)^2 ds$$
$$\leq Ch^2 \sum_{m=1}^{\infty} (v,\varphi_m)^2 \leq Ch^2\|v\|^2,$$

and, similarly,

$$\int_0^t s^2\|\rho_t\|^2 ds \le Ch^2\int_0^t s^2\|u_t(s)\|^2_{\dot{H}^1} ds \le Ch^2\|v\|^2,$$

so that

$$\|e(t)\|^2 \le Ch^2t^{-1}\|v\|^2.$$

To prove (2.6) in the general case, it suffices, by the stability of $E_h(t)$ and $E(t)$, to consider $ht^{-1/2} \le 1$. We now note that, using the semigroup property for $E_h(t)$ and $E(t)$, we may write

$$F_h(t) = F_h(t/2)E(t/2) + E(t/2)F_h(t/2) + F_h(t/2)^2$$
$$= G_1 + G_2 + G_3.$$

We have, by Theorem 2.1 and Lemma 2.2,

$$\|G_1(t)v\| \le Ch^r\|E(t/2)v\|_{\dot{H}^r} \le Ch^rt^{-r/2}\|v\|.$$

Further, since $E(t)$ and $E_h(t)$ are selfadjoint, we have $G_2 = G_1^*$, and hence G_2 has the same operator norm as G_1. By the result for r=1 we obtain

$$\|G_3(t)v\| \le Cht^{-1/2}\|F_h(t/2)v\|,$$

so that, altogether,

$$\|F_h(t)v\| \le Ch^rt^{-r/2}\|v\| + Cht^{-1/2}\|F_h(t/2)v\|.$$

Repeated application, using $ht^{-1/2} \le 1$, yields

$$\|F_h(t)v\| \le Ch^rt^{-r/2}\|v\| + C(ht^{-1/2})^r\|F_h(t/2^r)v\|$$
$$\le Ch^rt^{-r/2}\|v\|,$$

thus completing the proof.

We shall round off this section by using our error estimates for the homogeneous equation to show that, in the case of the nonhomogeneous equation, and for time bounded away from zero, regularity is required for optimal order convergence only near the time at which the error estimate is desired.

<u>Theorem 2.3</u>. Let u_h and u be the solution of (1.4) and (1.1), respectively, and let $v_h = P_h v$. Then, for $t \geq \delta > 0$, we have

$$\|u_h(t) - u(t)\| \leq Ch^r\{\|v\| + \int_0^t \|f\| ds + \int_{t-\delta}^t (\|u\|_r + \|u_t\|_r) ds\}.$$

<u>Proof</u>. We consider a fixed $t_0 \geq \delta$. Let $\varphi \in C^\infty$ be such that $\varphi(t) = 1$ for $t \geq -\delta/2$, $\varphi(t) = 0$ for $t \leq -\delta$, and set $\varphi_0(t) = \varphi(t-t_0)$. We now write $u = u_1+u_2+u_3$, where $u_1 = u\varphi_0$ and u_2 is the solution of the homogeneous equation

$$u_{2,t} - \Delta u_2 = 0, \quad t \geq 0,$$
$$u_2(0) = v.$$

Since

$$u_{1,t} - \Delta u_1 = f_1 \equiv f\varphi_0 + u\varphi_0', \quad \text{for } t \geq 0,$$
$$u_1(0) = 0,$$

it follows that u_3 satisfies

$$u_{3,t} - \Delta u_3 = f_3 \equiv f(1-\varphi_0) - u\varphi_0', \quad \text{for } t \geq 0,$$
$$u_3(0) = 0,$$

We note that f_1 and f_3 vanish for $t \leq t_0-\delta$ and $t \geq t_0-\delta/2$, respectively.

Let $u_{j,h}$, $j=1,2,3$, be the corresponding semidiscrete solutions with $u_{1,h}(\omega) = u_{3,h}(0) = 0$ and $u_{2,h}(0) = P_h v$, and set $e_j = u_{j,h} - u_j$. Since, by linearity, the total error at t_0 is the sum of the $e_j(t_0)$, it suffices to bound these quantities separately.

We obtain, by Theorem 1.1,

$$\|e_1(t_0)\| \le Ch^r\int_0^{t_0} \|u_{1,t}\|_r ds \le Ch^r\int_{t_0-\delta}^{t_0} (\|u\|_r+\|u_t\|_r)ds.$$

Further, by Theorem 2.2,

$$\|e_2\| \le Ct_0^{-r/2}h^r\|v\| \le C(\delta)h^r\|v\|.$$

Finally, by superposition, we have, for $t \ge t_0-\delta/2$

$$e_3(t) = \int_0^t F_h(t-s)f_3(s)ds = \int_0^{t_0-\delta/2} F_h(t-s)f_3(s)ds,$$

and hence

$$\|e_3(t_0)\| \le \int_0^{t_0-\delta/2} \|f_3\|ds \le Ch^r\int_0^{t_0} (\|f\| + \|u\|)ds$$
$$\le Ch^r(\|v\| + \int_0^{t_0} \|f\|ds).$$

Here the last step follows from the easily proven fact that

$$\|u(t)\| \le \|v\| + \int_0^t \|f\|ds, \quad \text{for } t \ge 0.$$

The proof is now complete.

The smooth and nonsmooth data results of Theorems 2.1 and 2.2 are from Bramble, Schatz, Thomée, and Wahlbin [4]. Results for nonsmooth data in the case of the homogeneous equations were first obtained by spectral methods in Blair [2], Thomée [38] and Helfrich [16], and later by energy methods in Sammon [35], Luskin and Rannacher [24], Huang and Thomée [17] and Thomée [40]. The result of Theorem 2.3 for the nonhomogeneous equation is from Thomée [39]. For a discussion of the effect of quadrature in the case of nonsmooth data, see Wahlbin [43]

3. COMPLETELY DISCRETE SCHEMES.

In this section we shall consider some completely discrete methods for the homogeneous parabolic equation

(3.1) $\quad u_t - \Delta u = 0, \quad \text{in } \Omega, \quad \text{for } t \geq 0,$

with initial and boundary conditions

(3.2) $\quad u = 0 \quad \text{on } \partial\Omega, \quad \text{for } t \geq 0,$

$\qquad u(\cdot,0) = v \quad \text{in } \Omega,$

where, as above, Ω is a bounded domain in R^d with smooth boundary $\partial\Omega$. We shall show some error estimates in the present context which are analogous to the ones presented in Section 2 above for the semidiscrete problem.

We shall thus assume, as above, that we have at our disposal a family of subspaces S_h of $H_0^1(\Omega)$ such that the standard approximation property (1.2) holds. As above we define the semidiscrete problem to find $u_h\colon [0,\infty) \to S_h$ such that

$$(u_{h,t},\chi) + (\nabla u_h,\nabla\chi) = 0, \qquad \forall\ \chi \in S_h, \quad t \geq 0,$$

$$u_h(0) = v_h \approx v.$$

As earlier this may be written in the form

(3.3) $\quad u_{h,t} - \Delta_h u_h = 0, \quad \text{for } t \geq 0,$

$\qquad u_h(0) = v_h,$

where

$$(\Delta_h\psi,\chi) = -(\nabla\psi,\nabla\chi) \qquad \forall\ \psi,\ \chi \in S_h.$$

Viewing the solution operator of this problem as the exponential $\exp(t\Delta_h)$, it is natural to define a completely discrete solution $U^n \in S_h$ at $t_n = nk$, where k is the time step, recursively by

$$(3.4)\qquad U^{n+1} = r(-k\Delta_h)U^n, \qquad n = 0,1\ldots$$
$$U^0 = v_h,$$

where $r(\lambda)$ is a rational function approximating the exponential $e^{-\lambda}$.

For instance, the backward Euler method,

$$(U^{n+1},\chi) + k(\nabla U^{n+1},\nabla\chi) = (U^n,\chi), \qquad \forall\ \chi \in S_h,\ n \geq 0,$$

corresponds to

$$r(\lambda) = \frac{1}{1+\lambda},$$

and the Crank-Nicolson method

$$(U^{n+1},\chi) + \tfrac{1}{2}k(\nabla U^{n+1},\nabla\chi) = (U^n,\chi) - \tfrac{1}{2}k(\nabla U^n,\nabla\chi),$$
$$\forall\ \chi \in S_h, \quad n \geq 0,$$

to

$$r(\lambda) = \frac{1 - \frac{1}{2}\lambda}{1 + \frac{1}{2}\lambda}.$$

We shall say that (3.4) is accurate of order p if

$$(3.5)\qquad r(\lambda) = e^{-\lambda} + O(\lambda^{p+1}), \quad \text{as} \quad \lambda \to 0.$$

The backward Euler method is then accurate of order 1, and the Crank-Nicolson method of order 2. More generally, the Padé approximants of $e^{-\lambda}$,

$$P_{\mu,\nu}(\lambda) = \frac{n_{\mu,\nu}(\lambda)}{d_{\mu,\nu}(\lambda)},$$

where

$$n_{\mu,\nu} = \sum_{j=0}^{\nu} \frac{(\mu+\nu-j)!\,\nu!}{(\mu+\nu)!\,j!\,(\nu-j)!}\,(-\lambda)^j$$

and

$$d_{\mu,\nu} = \sum_{j=0}^{\mu} \frac{(\mu+\nu-j)!\mu!}{(\mu+\nu)!j!(\mu-j)!} \lambda^j ,$$

satisfy (3.5) with $p = \mu+\nu$.

The relation (3.4) may be expressed in terms of the eigenvalues $\{\Lambda_j\}_1^{N_h}$ and eigenfunctions $\{\phi_j\}_1^{N_h}$ of the positive definite operator $-\Delta_h$: For $U^n \in S_h$ we have

$$U^n = \sum_{j=1}^{N_h} (U^n,\phi_j)\phi_j$$

and

$$U^n = r(-k\Delta_h)^n v_h = \sum_{j=1}^{N_h} r(k\Lambda_j)^n (v_h,\phi_j)\phi_j .$$

We note, in particular, that the method (3.4) is stable in L_2 if

(3.6) $$\max_j |r(k\Lambda_j)| \leq 1,$$

because then

$$\|U^n\|^2 = \sum_{j=1}^{N_h} |r(k\Lambda_j)^n|^2 (v_h,\phi_j)^2 \leq \sum_{j=1}^{N_h} (v_h,\phi_j)^2 = \|v_h\|^2 .$$

This condition is satisfied, in particular, for the backward Euler and Crank-Nicolson methods, and, more generally, for the diagonal and below diagonal Padé approximants, i.e. those with $\mu \leq \nu$.

We shall begin our error analysis by considering a non-smooth data estimate in which we make a considerably stronger assumption than (3.6), namely

(3.7) $$|r(\lambda)| < 1 \quad \text{for} \quad \lambda > 0, \qquad \text{with } |r(\infty)| < 1.$$

This condition is satisfied for the below diagonal Padé approximants, e.g. for the backward Euler method. Another example is the so called Calahan method, which corresponds to

$$r(\lambda) = 1 - \frac{\lambda}{1+b\lambda} - \frac{\sqrt{3}}{6}\left(\frac{\lambda}{1+b\lambda}\right)^2, \quad \text{with } b = \frac{1}{2}\left(1+\frac{\sqrt{3}}{3}\right).$$

Since this rational function has for its denominator the square of a linear function, the determination of U^{n+1} from U^n is accomplished by the solution of two consecutive linear systems with the same matrix.

We then have the following result.

Theorem 3.1. Let U^n and u_h be the solutions of (3.4) and (3.3), respectively. Assume that the method (3.4) is accurate of order p, that (3.7) holds, and that $v_h = P_h v$. Then we have, for the error in the time discretization of the semidiscrete problem (3.3),

$$\|U^n - u_h(t_n)\| \leq Ck^p t_n^{-p} \|v\|, \quad \text{for } t_n = nk > 0.$$

Proof. We may write

$$U^n - u_h(t_n) = r(-kA_n)^n P_h v - \exp(nkA_n) P_h v = F_n(-kA_n) P_h v,$$

where

$$F_n(\lambda) = r(\lambda)^n - e^{-n\lambda}.$$

Note that, therefore, by eigenfunction expansion,

$$U^n - u_h(t_n) = \sum_{j=1}^{N_h} F_n(k\Lambda_j)(v,\phi_j)\phi_j$$

and hence

$$\|U^n - u_h(t_n)\| \leq \max_j |F_n(k\Lambda_j)| \cdot \|v\|.$$

Consequently, it suffices to show

$$|F_n(\lambda)| \leq Ck^p t_n^{-p} = Cn^{-p}, \qquad \text{for } \lambda > 0.$$

We first prove this inequality for $\lambda \le \lambda_0$, say. By (3.5) it is clear that, for a suitable $c \in (0,1)$,

$$|r(\lambda)| \le e^{-c\lambda}, \quad \text{for} \quad 0 < \lambda \le \lambda_0.$$

Hence, for these λ,

$$|F_n(\lambda)| = |(r(\lambda) - e^{-\lambda}) \sum_{j=0}^{n-1} r(\lambda)^{n-1-j} e^{-j\lambda}|$$
$$\le Cn\lambda^{p+1} e^{-c(n-1)\lambda} \le Cn^{-p}(n\lambda)^{p+1} e^{-cn\lambda} \le Cn^{-p}.$$

We now consider $\lambda > \lambda_0$, and note that then

$$e^{-n\lambda} \le e^{-n\lambda_0} \le Cn^{-p}.$$

Further, since $|r(\infty)| < 1$, we also have by (3.7)

$$\sup_{\lambda \ge \lambda_0} |r(\lambda)| = e^{-c}, \quad \text{with} \quad c > 0,$$

so that

$$\sup_{\lambda \ge \lambda_0} |r(\lambda)^n| \le e^{-cn} \le Cn^{-p}.$$

Hence

$$|F_n(\lambda)| \le Cn^{-p}, \quad \text{for} \quad \lambda > 1,$$

which completes the proof.

Combining Theorem 3.1 with our previous semi-discrete non-smooth data error estimate, Theorem 2.1, we conclude at once the following.

Theorem 3.2. Under the assumptions of Theorem 3.1, and with u the solution of (3.1), (3.2), we have, for the total error in the completely discrete scheme (3.4),

$$\|U^n - u(t_n)\| \le C(h^r t_n^{-r/2} + k^p t_n^{-p})\|v\|, \quad \text{for } t_n = nk > 0.$$

Some stable time discretization methods, such as the Crank-Nicolson method, do not satisfy the requirement $|r(\infty)| < 1$ in (3.7), and the result of Theorem 3.1 is then not valid, as is easily seen by examining the proof. In such cases the conclusion is still valid provided the numbers $k\Lambda_j$ are uniformly bounded,

(3.8) $\quad k\Lambda_j \le \lambda_0, \qquad \text{for} \quad j = 1,\cdots,N_h.$

Such a condition is satisfied, with $\lambda_0 = \kappa_0\kappa_1$, if

(3.9) $\quad \Lambda_j \le \kappa_0 h^{-2}, \qquad \text{for } j = 1,\cdots,N_h,$

and if, in addition, the mesh-ratio restriction

(3.10) $\quad k/h^2 \le \kappa_1$

is imposed. In the situation described in Section 1, the condition (3.9) is a consequence of the inverse estimate

$$\|\nabla\chi\| \le c_0 h^{-1}\|\chi\|, \qquad \text{for} \quad \chi \in S_h,$$

since

$$\Lambda_j = -(\Delta_h\phi_j,\phi_j) = \|\nabla\phi_j\|^2 \le c_0^2 h^{-2}\|\phi_j\|^2 = \kappa_0 h^{-2}.$$

Both the conditions (3.9) and (3.10) are undesirable in practice. However, if they are satisfied, we have the following result.

<u>Theorem 3.3</u>. Assume that (3.5) holds, that

$$|r(\lambda)| < 1, \qquad \text{for } \lambda > 0,$$

and that (3.8) is valid. Then, if $v_h = P_h v$, we have

$$\|U^n - u_h(t_n)\| \le Ck^p t_n^{-p}\|v\|.$$

Proof. It follows from the proof of Theorem 3.1 that it suffices to show

$$|F_n(k\Lambda_j)| \le Cn^{-p}, \qquad \text{for } j = 1,\cdots,N_h,$$

or, in view of (3.8),

$$|F_n(\lambda)| \le Cn^{-p}, \qquad \text{for } 0 < \lambda \le \lambda_0.$$

But this was already shown in the proof of Theorem 3.1, which thus proves the theorem.

It was discovered by Rannacher that another way of securing the nonsmooth data estimate of Theorem 3.1, in the case of the diagonal Padé schemes, is to start with a few steps of a corresponding below diagonal scheme. We shall demonstrate this for the Crank-Nicolson scheme, starting with two steps of the backward Euler scheme, thus defining U^n by

$$(3.11)\qquad \begin{aligned} U^{n+1} &= r_1(-k\Delta_h)U^n, \text{ with } r_1(\lambda) = \frac{1-\lambda/2}{1+\lambda/2}, \text{ for } n \ge 2,\\ U^{n+1} &= r_0(-k\Delta_h)U^n, \text{ with } r_0(\lambda) = \frac{1}{1+\lambda}, \text{ for } n = 0,\ 1,\\ U^0 &= v_h. \end{aligned}$$

We then have the following result.

Theorem 3.4. With $v_h = P_h v$ we have, for the scheme (3.11),

$$\|U^n - u_h(t_n)\| \le Ck^2t_n^{-2}\|v\|, \qquad \text{for } n \ge 2.$$

Proof. In the same way as in the proof of Theorem 3.1 it suffices to show

$$|F_n(\lambda)| = |r_0(\lambda)^2 r_1(\lambda)^{n-2} - e^{-n\lambda}| \le Cn^{-2}, \qquad \text{for } \lambda > 0,\ n \ge 2,$$

and since both terms are bounded we may consider $n > 2$.

For large λ, $\lambda \geq \lambda_0$, say, we have with c suitable

$$|r_1(\lambda)| = \frac{1-2/\lambda}{1+2/\lambda} \leq e^{-c/\lambda}.$$

Hence

$$|r_0(\lambda)^2 r_1(\lambda)^{n-2}| \leq \frac{C}{\lambda^2} e^{-c(n-2)/\lambda}$$

$$\leq \frac{C}{(n-2)^2} \left(\frac{n-2}{\lambda}\right)^2 e^{-c(n-2)/\lambda} \leq \frac{C}{(n-2)^2} \leq \frac{C}{n^2}.$$

It follws that

$$|F_n(\lambda)| \leq Cn^{-2} + Ce^{-n\lambda_0} \leq Cn^{-2}, \text{ for } \lambda > \lambda_0.$$

To consider $\lambda \leq \lambda_0$, we write

$$F_n(\lambda) = r_0^2(r_1^{n-2} - e^{-(n-2)\lambda}) + (r_0^2 - e^{-2\lambda})e^{-(n-2)\lambda}.$$

By the argument of the proof of Theorem 3.1 we have, for $\lambda \leq \lambda_0$,

$$|r_1(\lambda)^{n-2} - e^{-(n-2)\lambda}| \leq \frac{C}{(n-2)^2} \leq Cn^{-2},$$

and

$$|r_0^2 - e^{-2\lambda}| \leq C\lambda^2,$$

so that

$$|F_n(\lambda)| \leq Cn^{-2} + C\lambda^2 e^{-n\lambda} \leq Cn^{-2}, \text{ for } \lambda \leq \lambda_0.$$

Together these estimates complete the proof.

The above analysis was developed in Baker, Bramble and Thomée [1], see also Sammon [36], Bramble and Sammon [3].

We shall briefly consider the possibility of using more than two time levels in our finite difference discretization in time. Denoting as earlier the backward difference quotient operator by $\bar{\partial}_t$ we may replace the time derivative in (3.3) by the most accurate approximation based on $U^n, U^{n-1}, \ldots, U^{n-q}$ to obtain

$$\sum_{j=1}^{q} \frac{k^{j-1}}{j} \bar{\partial}_t^j U^n - \Delta_h U^n = 0, \quad \text{for} \quad n \geq q.$$

In addition to setting $U^0 = v_h$, this equation then has to be combined with some other starting procedure to determine $U^1, \ldots, U^{q-1}$.

We shall consider in some detail the second order accurate case of $q = 2$, with U^1 defined by the backward Euler method,

$$(3.12) \qquad \begin{aligned} &(\tfrac{3}{2} U^n - 2U^{n-1} + \tfrac{1}{2} U^{n-2})/k - \Delta_h U^n = 0, \qquad n \geq 2, \\ &(U^1 - U^0)/k - \Delta_h U^1 = 0, \\ &U^0 = v_h. \end{aligned}$$

We shall show the following.

<u>Theorem 3.5</u>. With $v_h = P_h v$ we have for the backward differencing method (3.12)

$$\|U^n - u_h(t_n)\| \leq Ck^2 t_n^{-2} \|v\|, \quad \text{for} \quad n \geq 2.$$

<u>Proof.</u> Setting, with our earlier notation,

$$U^n = \sum_{j=1}^{N_h} \gamma_j^n \phi_j,$$

we find for the γ_j^n the difference equation

$$(\tfrac{3}{2} + k\Lambda_j)\gamma_j^n - 2\gamma_j^{n-1} + \tfrac{1}{2} \gamma_j^{n-2} = 0, \quad n \geq 2,$$

with the initial conditions

$$\gamma_j^0 = v_j = (v, \phi_j),$$
$$\gamma_j^1 = (1+k\Lambda_j)^{-1} v_j = r_0(k\Lambda_j) v_j, \quad \text{where} \quad r_0(\lambda) = \frac{1}{1+\lambda}.$$

The general solution of the difference equation is associated with the quadratic equation

$$(\frac{3}{2} + \lambda)\rho^2 - 2\rho + \frac{1}{2} = 0,$$

with roots

$$r_{1,2}(\lambda) = \frac{2\pm\sqrt{1-2\lambda}}{3+2\lambda}, \qquad 0 < \lambda \leq 1/2.$$

$$= \frac{2\pm i\sqrt{2\lambda-1}}{3+2\lambda}, \qquad \lambda \geq 1/2,$$

where for $\lambda < 1/2$ we let r_1 denote the larger of the two. For $\lambda \neq 1/2$ the roots are distinct and thus for $k\Lambda_j \neq 1/2$,

$$\gamma_j^n = [c_1(k\Lambda_j)r_1(k\Lambda_j)^n + c_2(k\Lambda_j)r_2(k\Lambda_j)^n]v_j, \quad n \geq 0,$$

where c_1 and c_2 are determined by the initial conditions as

$$c_1 = c_1(\lambda) = \frac{r_0 - r_2}{r_1 - r_2},$$

$$c_2 = c_2(\lambda) = \frac{r_1 - r_0}{r_1 - r_2}.$$

For $k\Lambda_j = 1/2$ we have similarly

$$\gamma_j^n = (1+nd(k\Lambda_j))r_1(k\Lambda_j)^n u_j$$

where

$$d = d(\lambda) = r_0(r_1 - 1).$$

Hence

$$U^n - u_h(t_n) = \sum_{j=1}^{N_h} F_n(k\Lambda_j)v_j\phi_j$$

where

$$F_n = F_n(\lambda) = \frac{r_0 - r_2}{r_1 - r_2} r_1^n + \frac{r_1 - r_0}{r_1 - r_2} r_2^n - e^{-n\lambda}, \quad \text{if} \quad \lambda \neq 1/2$$

$$= nr_0r_1^{n-1} - (n-1)r_1^n - e^{-n\lambda}, \qquad \text{if} \quad \lambda = 1/2.$$

The desired result now follows if we can prove

$$|F_n(\lambda)| \leq Cn^{-2}, \quad \text{for} \quad \lambda > 0, \tag{3.13}$$

since then

$$\|U^n - u_h(t_n)\|^2 = \sum_{j=1}^{N_h} F_n(k\Lambda_j)^2 v_j^2$$

$$\leq Cn^{-2}\|v\|^2 = Ck^2 t_n^{-2}\|v\|^2.$$

For λ small we have

$$r_1(\lambda) = 1 - \lambda + \frac{1}{2}\lambda^2 + O(\lambda^3) = e^{-\lambda} + O(\lambda^3),$$

and, in particular, $\delta > 0$ suitable,

$$|r_1(\lambda)| \leq e^{-\lambda/2}, \quad \text{for} \quad 0 < \lambda < \delta.$$

Further,

$$c_1(\lambda)r_1(\lambda)^n - e^{-n\lambda} = r_1(\lambda)^n - e^{-n\lambda} + (c_1(\lambda) - 1)r_1(\lambda)^n$$

$$= O(n^{-2} + \lambda^2 e^{-n\lambda/2}) = O(n^{-2}), \quad \text{as} \quad n \to \infty.$$

Since r_2 is bounded away from 1 and c_2 is bounded this shows (3.13) for $\lambda < \delta$.

For $\lambda > \delta$, $\lambda \neq 1/2$, we may write

$$F_n = r_0 \frac{r_1^n - r_2^n}{r_1 - r_2} + r_1^n + r_2^n - \frac{r_1^{n+1} - r_2^{n+1}}{r_1 - r_2} - e^{-n\lambda},$$

which is bounded by $c\gamma^{n-1}$ with $\gamma < 1$ since r_1 and r_2 are uniformly smaller than 1, and for $\lambda = 1/2$ the corresponding estimate is obtained by passing to the limit. This implies (3.13) for $\lambda > \delta$, and thus completes the proof of the theorem.

It is clear that this result may also be combined with our estimate for the semidiscrete problem to yield an estimate like that of Theorm 3.2 with $p=2$. Similar results told for the other backward differencing method with $q \leq 6$, in which case the order of accuracy is $p = q$.

For further results concerning multistep methods, see Zlámal [46], Le Roux [21], [22].

We shall briefly turn to a renewed discussion of the error in the discretization in time of the semidiscrete problem (3.3). Since the error bound in Theorem 3.1 is large for small t it appears natural to try to obtain a more uniform error bound by taking smaller time steps in the beginning of the computation. Following Johnson, Nie, and Thomée [19] we shall analyze such a procedure for the backward Euler method.

Let thus $0 = t_0 < t_1 < \cdots < t_n < \cdots$ be a partition of the positive time axis and set $I_n = (t_{n-1}, t_n)$ and $k_n = t_n - t_{n-1}$. We shall then consider the approximation U^n of $u^n = u(t_n)$ defined by

$$(3.14) \qquad \frac{U^n - U^{n-1}}{k_n} - \Delta_h U^n = 0, \quad \text{for} \quad n \geq 1,$$
$$U^0 = v_h,$$

or

$$U^n = E_{k_n} U^{n-1}, \quad \text{for } n \geq 1, \quad \text{with } E_k = (I - k\Delta_h)^{-1},$$
$$U^0 = v_h,$$

or, in a concise form

$$U^n = E_{n,1} v_h,$$

where we have introduced the notation

$$E_{n,j} = E_{k_n} E_{k_{n-1}} \cdots E_{k_j}, \quad \text{for} \quad j \leq n.$$

The error $\eta^n = U^n - u_h^n$, then satisfies

$$(3.15) \quad \frac{\eta^n - \eta^{n-1}}{k_n} - \Delta_h \eta^n = \omega^n = -\frac{u_h^n - u_h^{n-1}}{k_n} + \Delta_h u_h^n = u_{h,t}^n - \frac{u_h^n - u_h^{n-1}}{k_n}.$$

Hence, we have

$$\eta^n = E_{k_n}\eta^{n-1} + k_n E_{k_n}\omega^n,$$

or, by repeated application, since $\eta_0 = 0$,

$$\eta^n = \sum_{j=1}^{n} k_j E_{n,j}\omega^j, \quad \text{for } n \geq 1. \tag{3.16}$$

As before, $\|E_k\| \leq 1$, so that $\|E_{n,j}\| \leq 1$, and thus

$$\|\eta^n\| \leq \sum_{j=1}^{n} k_j \|\omega^j\|.$$

Since

$$\|\omega^n\| = \|u^n_{h,t} - \frac{u^n_h - u^{n-1}_h}{k_n}\| \leq C\int_{I_j} \|u_{h,tt}\| dt,$$

we conclude that

$$\|\eta^n\| \leq C\sum_{j=1}^{n} k_j \int_{I_j} \|u_{h,tt}\| dt. \tag{3.17}$$

For a uniform partitioning this reduces to the estimate

$$\|\eta^n\| \leq Ck\int_0^{t_n} \|u_{h,tt}\| dt,$$

which is similar in form to the contribution to the discretization in time in the result of Theorem 1.3.

We shall now present an alternative error bound to that of (3.17) in which the sum is replaced by a maximum and where only the first order derivative of u_h with respect to time enters.

Theorem 3.6. We have for the variable time step method (3.14)

$$\|U^n - u_h(t_n)\| \leq C(1 + \log \frac{t_n}{k_n}) \max_{j\leq n} \int_{I_j} \|u_{h,t}\| \, dt.$$

Proof. We write (3.16) in the form

$$\eta^n = \sum_{j=1}^{n} k_j E_{n,j}(-\Delta_h)((-\Delta_h)^{-1}\omega^j).$$

Our result will follow from

$$\sum_{j=1}^{n} k_j \|E_{n,j}(-\Delta_h)\| \le 1 + \log \frac{t_n}{k_n}$$

and

$$\|(-\Delta_h)^{-1}\omega^j\| \le C\int_{I_j} \|u_{h,t}\|\, ds. \tag{3.18}$$

To show the first of these two inequalities we note that, by spectral representation,

$$\|E_{n,j}(-\Delta_h)\| \le \max_{\lambda\ge 0} \frac{\lambda}{(1+k_n\lambda)\cdot\cdots\cdot(1+k_j\lambda)}$$
$$\le \frac{1}{k_j+\cdots+k_n} = \frac{1}{t_n-t_{j-1}}.$$

Hence

$$\sum_{j=1}^{n} k_j\|E_{n,j}(-\Delta_h)\| \le \sum_{j=1}^{n} \frac{k_j}{t_n-t_{j-1}}$$
$$\le 1 + \sum_{j=1}^{n-1}\int_{I_j} \frac{ds}{t_n-s} = 1 + \log\frac{t_n}{k_n}.$$

We have from (3.15)

$$\omega^j = -\frac{1}{k_j}\int_{I_j} u_{h,t}ds + \Delta_h u_h^j = -\frac{1}{k_j}\int_{I_j} \Delta_h u_h ds + \Delta_h u_h^j$$
$$= -\Delta_h\{\frac{1}{k_j}\int_{I_j} u_h dt - u_h^j\},$$

from which (3.18) follows at once. This completes the proof.

The above result is valid also for the discretization in time of the nonhomogeneous semidiscrete equation

$$u_{h,t} - \Delta_h u_h = f_h \equiv P_h f,$$

provided we define the U^n from the modified backward Euler equation

$$(3.19)\qquad \frac{U^n-U^{n-1}}{k_n} - \Delta_h U^n = \frac{1}{k_n}\int_{I_n} f_h ds.$$

The standard backward Euler method may be thought of as obtained from (3.19) by applying quadrature to the right hand side. In the present form the method may also be expressed as the result of a Galerkin procedure in time, using discontinuous, piecewise constant approximations in the time variable, cf. [19].

Since, for most practical choices of the time steps, the logarithmic factor is of moderate size one may use the result of the theorem, at least in theory, to bound the error essentially uniformly in time by choosing the k_j such that $\int_{I_j} \|u_{h,t}\| dt$ is kept uniformly small. This may be accomplished by choosing k_j such that $k_j \max_{I_j} \|u_{h,t}\|$ is constantly small.

For example, assume $v_h = R_h v$ with $v \in H_0^1(\Omega)$. Then the standard spectral argument shows

$$\|u_{h,t}(t)\| \le Ct^{-1/2}\|\nabla R_h v\| \le Ct^{-1/2}\|\nabla v\| = C_0 t^{-1/2},$$

and hence

$$\int_{I_j} \|u_{h,t}\| dt \le 2C_0 k_1^{1/2}, \quad \text{for } j = 1,$$

$$\le C_0 t_{j-1}^{-1/2} k_j, \quad \text{for } j > 1.$$

This suggests choosing $k_1 = (\delta/(2C_0))^2$, and then, for $j>1$, $k_j = (\delta/C_0) t_{j-1}^{1/2}$. In such a way the error will be bounded by δ, uniformly for $n\ge1$, modulo a logarithmic factor.

The difficulty with the procedure proposed, particularly for the nonhomogeneous equation, is that u_h is not known and hence the $\int_{I_j} \|u_{h,t}\| dt$ also not. However, in some cases these quantities may be estimated from the computed solution and then be used for an adaptive time stepping method. We refer to [19] for details.

We shall close by considering briefly error estimates which hold uniformly down to t=0. Here regularity has to be assumed from the initial data, which will be expressed in terms of the spaces $\dot{H}^s(\Omega)$ introduced in Section 2. We recall that for the semidiscrete problem, with the appropriate choice of discrete initial values v_h, we have

(3.20) $\quad \|u_h(t) - u(t)\| \leq Ch^r \|v\|_{\dot{H}^r}, \quad$ for $t \geq 0$.

In order to obtain a similar result for a completely discrete method we shall combine (3.20) with some estimates derived by the technique used in the proof of Theorem 3.1 and with the easily verified identity

(3.21) $\quad v = \sum_{j=0}^{p-1} T_h^j (I-R_h)(-\Delta)^j v + T_h^p (-\Delta)^p v, \quad$ for $v \in \dot{H}^{2p}(\Omega)$.

The following is our result in this case (cf. Baker, Bramble and Thomée [1]).

Theorem 3.7. Assume that the approximation property (1.2) holds, that the scheme is of order p in time, and that

$$|r(\lambda)| < 1, \qquad \text{for } \lambda > 0.$$

Let $v \in \dot{H}^{\max(r,2p)}(\Omega)$ and let v_h be chosen so that

$$\|v_h - v\| \leq Ch^r \|v\|_{\dot{H}^r}.$$

Then

$$\|U^n - u(t_n)\| \leq C\{h^r \|v\|_{\dot{H}^r} + k^p \|v\|_{\dot{H}^{2p}}\}, \qquad \text{for } t \geq 0.$$

Proof. We shall show that, with the notation used earlier,

$$(3.22) \qquad \|F_n(-k\Delta_h)T_h^j\| \leq Ck^j, \text{ for } 0 \leq j \leq p,\ n \geq 0.$$

Assuming this for a moment we shall prove our result, and note first that it suffices to consider the case $v_h = P_h v$, because, by stability,

$$\|r(-k\Delta_h)^n(v_h - P_h v)\| \leq \|v_h - v\| + \|P_h v - v\| \leq Ch^r \|v\|_{\dot{H}^r}.$$

We may now write, with $\{\lambda_m\}$ and $\{\varphi_m\}$ the eigenvalues and eigenfunctions of $-\Delta$,

$$U^n - u_h(t_n) = F_n(-k\Delta_h)v.$$

We define

$$v_k = \sum_{k\lambda_m \leq 1} (v, \varphi_k)\varphi_k,$$

and note that our definitions imply

$$(3.23) \qquad \|v_k\|_{\dot{H}^{2p}} \leq \|v\|_{\dot{H}^{2p}},$$

$$(3.24) \qquad \|v_k\|_{\dot{H}^{r+2j}} \leq Ck^{-j}\|v\|_{\dot{H}^r}, \qquad j = 0,\ldots,p-1,$$

and

$$(3.25) \qquad \|v - v_k\| \leq Ck^p \|v\|_{\dot{H}^{2p}}.$$

Now, with $F_n = F_n(-k\Delta_h)$, we have by (3.21)

$$F_n v_k = \sum_{j=0}^{p-1} F_n T_h^j (I-R_h)(-\Delta)^j v_k + F_n T_h^p (-\Delta)^p v_k.$$

Here, by (3.22) and (3.23),

$$\|F_n T_h^p (-\Delta)^p v_k\| \le Ck^p \|\Delta^p v_k\| \le Ck^p \|v_k\|_{\dot{H}^{2p}} \le Ck^p \|v\|_{\dot{H}^{2p}},$$

and, using now also (3.24),

$$\|F_n T_h^j (I-R_h)(-\Delta)^j v_k\| \le Ck^j \|(I-R_h)(-\Delta)^j v_k\|$$
$$\le Ck^j h^r \|v_k\|_{\dot{H}^{r+2j}} \le Ch^r \|v\|_{\dot{H}^r},$$

and finally, with (3.25)

$$\|F_n(v - v_k)\| \le C\|v - v_k\| \le Ck^p \|v\|_{\dot{H}^{2p}}.$$

It remains to show (3.22). By eigenfunction expansion this follows from

$$|F_n(\lambda)| \le C\lambda^j, \qquad \text{for } \lambda \ge 0,$$

and it is clear that it suffices to show this inequality for $0 \le \lambda \le \lambda_0$ with $\lambda_0 > 0$. But in this case it follows at once from (3.5). This completes the proof.

In this section we have only considered the discretization in time of a homogeneous parabolic equation. For formulation and analysis of completely discrete methods for nonhomogeneous equations, see Crouzeix [8], Brenner, Crouzeix, and Thomée [5], Thomée [40], Eriksson, Johnson, and Thomée [13].

4. A NONLINEAR PROBLEM.

In this section we shall apply our previous analysis to a nonlinear problem. For simplicity and concreteness, we mainly restrict our attention to the situation in the beginning of Section 1 with two space dimensions and with piecewise linear approximating functions.

Let thus Ω be a plane convex domain with smooth boundary and consider the parabolic problem

$$(4.1)\qquad \begin{aligned} &u_t - \nabla\cdot(a(u)\nabla u) = f(u), \quad \text{in } \Omega, \text{ for } t \in J = [0,T], \\ &u = 0, \quad \text{on } \partial\Omega, \text{ for } t \in J, \\ &u(\cdot,0) = v, \quad \text{in } \Omega, \end{aligned}$$

where a and f are smooth functions with a positive,

$$(4.2)\qquad 0 < \mu \leq a(u) \leq M, \quad \text{for } u \in R,$$

and with both a and f globally Lipschitz continuous, so that for $u, v \in R$,

$$(4.3)\qquad \begin{aligned} &|a(u) - a(v)| \leq L|u - v|, \\ &|f(u) - f(v)| \leq L|u - v|. \end{aligned}$$

This time, we are thus restricting the considerations to a finite interval $J = [0,T]$ in time, and some of the constants below will depend on T. We assume that the above problem admits a unique solution, which is sufficiently smooth for our purposes.

Let now, as in Section 1, $\mathcal{T}_h$ be a triangulation of Ω with $\max_{\tau\in\mathcal{T}_h} \text{diam}\,\tau \leq h$, and let S_h be the corresponding finite dimensional space of continuous functions on Ω, which reduce to linear functions in each of the triangles of $\mathcal{T}_h$ and which vanish on $\partial\Omega$. We may then pose the semidiscrete problem to find $u_h: J \rightarrow S_h$ such that

(4.4) $$(u_{h,t}, \chi) + (a(u_h)\nabla u_h, \nabla\chi) = (f(u_h), \chi), \qquad \forall\, \chi \in S_h,$$
$$u_h(0) = v_h,$$

where again v_h is an approximation of v in S_h. Representing the solution in the form

$$u_h(x,t) = \sum_{j=1}^{N_h} \alpha_j(t)\phi_j(x),$$

where $\{\phi_j\}_1^{N_h}$ is the standard basis of pyramid functons, this may be written

$$\sum_{j=1}^{N_h} \alpha_j'(t)(\phi_j, \phi_k) + \sum_{j=1}^{N_h} \alpha_j(t)(a(\sum_{\ell=1}^{N_h} \alpha_\ell(t)\phi_\ell)\nabla\phi_j, \nabla\phi_k)$$
$$= (f(\sum_{\ell=1}^{N_h} \alpha_\ell(t)\phi_\ell), \phi_k), \qquad k = 1,\cdots,N_h.$$

Setting $\alpha(t) = (\alpha_1(t),\cdots,\alpha_{N_h}(t))^T$ and $A = (a_{jk})$ with $a_{jk} = (\phi_j, \phi_k)$, this may also be stated in the form

(4.5) $$A\alpha'(t) = F(\alpha(t)), \qquad \text{for} \quad t \in J,$$
$$\alpha(0) = \gamma.$$

Since A is positive definite, this nonlinear system of ordinary differential equations has a unique solution, at least locally. In fact, because of our assumptions about a and f, the vector valued function F is globally Lipschitz continuous, and the solution $\alpha(t)$ exists for all positive t.

We wish to estimate the error in the semidiscrete problem. As earlier we shall write the error

(4.6) $$u_h - u = (u_h - \tilde{u}_h) + (\tilde{u}_h - u) = \theta + \rho,$$

where $\tilde{u}_h$ is an elliptic projection in S_h of the exact solution u. This time we shall use the projection defined by

(4.7) $$(a(u)\nabla(\tilde{u}_h - u), \nabla\chi) = 0, \qquad \forall\, \chi \in S_h,$$

and we shall therefore need some estimates for the error in this projection.

<u>Lemma 4.1</u>. Let a be a smooth function in Ω with

$$0 < \mu \leq a(x) \leq M, \qquad \text{for } x \in \Omega,$$

and let v_h be defined by

$$(a\nabla(v_h - v), \nabla\chi) = 0, \qquad \forall\, \chi \in S_h.$$

Then

(4.8) $$\|\nabla(v_h - v)\| \leq C_1 h \|v\|_2$$

and

(4.9) $$\|v_h - v\| \leq C_0 h^2 \|v\|_2,$$

where C_1 depends on the triangulation $\mathcal{T}_h$ and μ and M, and C_0, in addition, on an upper bound for ∇a.

<u>Proof</u>. We have for $\chi \in S_h$

$$\mu\|\nabla(v_h - v)\|^2 \leq (a\nabla(v_h - v), \nabla(v_h - v))$$

$$= (a\nabla(v_h - v), \nabla(\chi - v)) \leq M\|\nabla(v_h - v)\| \cdot \|\nabla(\chi - v)\|,$$

and hence, with $I_h v$ the interpolant of v,

$$\|\nabla(v_h - v)\| \leq \|\nabla(I_h v - v)\| \leq Ch\|v\|_2,$$

which is (4.8). To show (4.9) by duality, we solve, with φ

arbitrary in $L_2(\Omega)$, the problem

$$(4.10)\qquad -\nabla\cdot(a(u)\nabla\psi) = -a\Delta\psi - \nabla a\cdot\nabla\psi = \varphi, \quad \text{in } \Omega,$$

$$\psi = 0, \quad \text{on } \partial\Omega,$$

and note that

$$\mu\|\nabla\psi\|^2 \le (a\nabla\psi,\nabla\psi) = (\varphi,\psi) \le \|\varphi\|\cdot\|\psi\| \le C\|\varphi\|\cdot\|\nabla\psi\|,$$

so that

$$\|\nabla\psi\| \le C\|\varphi\|,$$

and hence, since ∇a is bounded,

$$\|\psi\|_2 \le C\|\Delta\psi\| \le C\|a\Delta\psi\| = C\|\varphi + \nabla a\cdot\nabla\psi\| \le C\|\varphi\|.$$

Therefore, now with $\chi = I_h\psi$,

$$\begin{aligned}(v_h-v,\varphi) &= (a\nabla(v_h-v),\nabla\psi) = (a\nabla(v_h-v),\nabla(\psi-\chi))\\ &\le M\|\nabla(v_h-v)\|\cdot\|\nabla(\psi-\chi)\| \le (Ch\|v\|_2)(Ch\|\psi\|_2)\\ &\le Ch^2\|v\|_2\|\varphi\|,\end{aligned}$$

which completes the proof.

Since $\nabla a(u) = a'(u)\nabla u$ we may conclude, under the appropriate regularity assumptions for u, the following estimate.

<u>Lemma 4.2</u>. With $\rho = \tilde{u}_h - u$, we have

$$\|\rho\| + h\|\nabla\rho\| \le C(u)h^2.$$

We shall also need an estimate for ρ_t.

<u>Lemma 4.3</u>. We have, for $\rho = \tilde{u}_h - u$,

$$\|\rho_t\| + h\|\nabla\rho_t\| \le C(u)h^2.$$

<u>Proof</u>. We begin with the estimate for the gradient. By differentiation of the equation (4.7) we have

$$(a(u)\nabla\rho_t,\nabla\chi) + (a(u)_t\nabla\rho,\nabla\chi) = 0, \quad \forall\ \chi \in S_h.$$

Hence

$$\begin{aligned}\mu\|\nabla\rho_t\|^2 &\le (a(u)\nabla\rho_t,\nabla\rho_t)\\ &= (a(u)\nabla\rho_t,\nabla(\chi-u_t)) + (a(u)\nabla\rho_t,\nabla(\tilde{u}_{h,t}-\chi))\\ &= (a(u)\nabla\rho_t,\nabla(\chi-u_t)) - (a(u)_t\nabla\rho,\nabla(\tilde{u}_{h,t}-\chi))\\ &\le C(\|\nabla\rho_t\|\cdot\|\nabla(\chi-u_t)\| + \|\nabla\rho\|\cdot\|\nabla(\tilde{u}_{h,t}-\chi)\|),\end{aligned}$$

and, with $\chi = I_h\rho_t$,

$$\begin{aligned}\mu\|\nabla\rho_t\|^2 &\le Ch\|u_t\|_2\|\nabla\rho_t\| + \|\nabla\rho\|(Ch\|u_t\|_2 + \|\nabla\rho_t\|)\\ &\le \frac{\mu}{2}\|\nabla\rho_t\|^2 + C(\|\nabla\rho\|^2 + h^2\|u_t\|_2^2).\end{aligned}$$

In view of Lemma 4.2, this shows

$$\|\nabla\rho_t\| \le C(u)h.$$

For the L_2 estimate we use again the duality argument of the proof of Lemma 4.1. We have, with φ and ψ as in (4.10), (with $a = a(u)$),

$$\begin{aligned}(\rho_t,\varphi) &= (a(u)\nabla\rho_t,\nabla\psi)\\ &= (a(u)\nabla\rho_t,\nabla(\psi-\chi)) + (a(u)_t\nabla\rho,\nabla(\psi-\chi)) - (\nabla\rho,a(u)_t\nabla\psi),\end{aligned}$$

and hence, choosing $\chi = I_h\psi$ and using integration by parts in the last term,

$$|(\rho_t,\varphi)| \le C(\|\nabla\rho_t\|\ h\|\psi\|_2 + \|\nabla\rho\|\ h\|\psi\|_2 + \|\rho\|\cdot\|\psi\|_2),$$

whence, by Lemma 4.2 and the above estimate for $\nabla\rho_t$,

$$|(\rho_t,\varphi)| \le Ch^2\|\varphi\|,$$

or

$$\|\rho_t\| \le C(u)h^2,$$

which completes the proof of the lemma.

We shall also need below the boundedness of $\nabla\tilde{u}_h$.

Lemma 4.4. We have independently of t and h,

$$\|\nabla\tilde{u}_h\|_{L_\infty} \le C(u).$$

Proof. We have

$$\|\nabla\tilde{u}_h\|_{L_\infty} \le \|\nabla(\tilde{u}_h - I_h u)\|_{L_\infty} + \|\nabla I_h u\|_{L_\infty}.$$

Using an inverse estimate together with Lemma 4.2 and the known error estimate for $I_h u$ we have

$$\|\nabla(\tilde{u}_h - I_h u)\|_{L_\infty} \le Ch^{-1}\|\nabla(\tilde{u}_h - I_h u)\|$$

$$\le Ch^{-1}(\|\rho\| + \|\nabla(I_h u - u)\|) \le C(u).$$

Since it is easy to see that

$$\|\nabla I_h u\|_{L_\infty} \le C\|\nabla u\|_{L_\infty},$$

the result follows.

We now turn to the L_2 error estimate for the parabolic problem.

Theorem 4.1. Under the appropriate regularity assumptions for u, we have for the error in the semidiscrete problem (4.4), with $C = C(u)$,

$$\|u_h(t) - u(t)\| \le C(\|v_h - v\| + h^2).$$

Proof. In view of (4.6) and Lemma 4.2, it suffices to bound $\theta = u_h - \tilde{u}_h$. We have

$$\begin{aligned}
&(\theta_t,\chi) + (a(u_h)\nabla\theta,\nabla\chi)\\
&= (u_{h,t},\chi) + (a(u_h)\nabla u_h,\nabla\chi) - (\tilde{u}_{h,t},\chi) - (a(u_h)\nabla\tilde{u}_h,\nabla\chi)\\
&= (f(u_h),\chi) - (\rho_t,\chi) - (u_t,\chi)\\
&\qquad - (a(u)\nabla\tilde{u}_h,\nabla\chi) + ((a(u)-a(u_h))\nabla\tilde{u}_h,\nabla\chi)\\
&= (f(u_h),\chi) - (\rho_t,\chi) - (u_t,\chi)\\
&\qquad - (a(u)\nabla u,\nabla\chi) + ((a(u)-a(u_h))\nabla\tilde{u}_h,\nabla\chi)\\
&= (f(u_h)-f(u),\chi) + ((a(u)-a(u_h))\nabla\tilde{u}_h,\nabla\chi) - (\rho_t,\chi),
\end{aligned}$$

or, with $\chi = \theta$, since $\nabla\tilde{u}_h$ is bounded by Lemma 4.4,

$$\begin{aligned}
\frac{1}{2}\frac{d}{dt}\|\theta\|^2 + \mu\|\nabla\theta\|^2 &\le C\{\|u_h-u\|(\|\theta\| + \|\nabla\theta\|) + \|\rho_t\|\cdot\|\theta\|\}\\
&\le C(\|\theta\| + \|\rho\| + \|\rho_t\|)\|\nabla\theta\|\\
&\le \mu\|\nabla\theta\|^2 + C(\|\theta\|^2 + \|\rho\|^2 + \|\rho_t\|^2).
\end{aligned}$$

Hence, using Grönwall's lemma,

$$\|\theta(t)\|^2 \le C\|\theta(0)\|^2 + C\int_0^t(\|\rho\|^2 + \|\rho_t\|^2)ds.$$

In view of Lemmas 4.2 and 4.3, this shows

$$\|\theta(t)\| \le C(\|v_h - v\| + h^2),$$

and thus completes the proof.

The corresponding estimate for the gradient follows easily by an inverse estimate. We refrain from giving the details.

We shall now pause to make a comment concerning the global nature of our assumptions (4.2) and (4.3) for the functions a and f. It should be clear from our analysis that as long as u_h is close to u, the assumptions referred to only come into play in a neighborhood of the range of u. It is therefore natural to make the less stringent assumption that a and f are defined in such a neighborhood and satisfy (4.2) and (4.3) there. It has to be kept in mind, however, that closeness now has to be interpreted as being valid for each point, or that u_h be close to u in the uniform norm.

Thus, let I_0 be the range of u,

$$I_0 = [a,b] = \{u(x,t);\ x \in \bar{\Omega},\quad t \in J\},$$

and consider for a fixed $\delta > 0$ the interval $I_\delta = [a-\delta, b+\delta]$. Assume now that f and a are defined and sufficiently smooth on I_δ, in particular, Lipschitz continuous there, and that a is positive and bounded away from 0 and ∞. Then, if v_h is sufficiently close to v, or $v_h(x) \in I_{\delta/2}$ for $x \in \Omega$, say, we have that the problem (4.4) or (4.5) is well defined and has a solution in I_δ, at least for t in an interval $[0,t_h]$ with $0 < t_h \leq T$. Assume, for instance, that v_h is chosen so that

$$\|v_h - v\| \leq Ch^2.$$

Then, by the standard inverse estimate, with $\chi = I_h v$,

$$\|v_h - v\|_{L_\infty} \leq \|v_h - \chi\|_{L_\infty} + \|\chi - v\|_{L_\infty}$$
$$\leq Ch^{-1}\|v_h - \chi\| + \|\chi - v\|_{L_\infty} \leq C(v)h,$$

so that $v_h \in I_{\delta/2}$ for small h. As long as $u_h(t) \in I_\delta$, however, the above error analysis remains valid, and we conclude

$$\|u_h(t) - u(t)\| \leq C(\|v_h - v\| + h^2) \leq C(u)h^2,$$

and thus again, for $t \leq t_h$,

$$\|u_h(t) - u(t)\|_{L_\infty} \leq C(u)h \leq \delta/2, \qquad \text{for } h \leq h_0,$$

where h_0 is independent of t_h. Thus, $u_h(t_h) \in I_{\delta/2}$, and the solution continues to exist beyond t_h. We now conclude that

t_h may be chosen as T and thus that our local assumptions for a and f suffice in the proof of Theorem 4.1, for h small. On the other hand, this also means that the local assumptions may be extended to hold globally so that (4.2) and (4.3) may be assumed without restriction of generality.

We shall now turn to completely discrete schemes and begin with the backward Euler-Galerkin scheme which in this case reads

$$(4.11)\qquad (\bar{\partial}_t U^n,\chi) + (a(U^n)\nabla U^n,\nabla\chi) = (f(U^n),\chi), \quad \forall\, \chi \in S_h, \quad n \geq 1,\ nk \leq T,$$

$$U^0 = v_h,$$

where, with k the time step, U^n is the approximation at time nk and $\bar{\partial}_t U^n = k^{-1}(U^n - U^{n-1})$.

Introducing the vector α^n by

$U^n = \sum_{j=1}^{N_h} \alpha_j^n \phi_j$, the equation (4.11) may be written in matrix form as

$$A\frac{\alpha^n - \alpha^{n-1}}{k} = F(\alpha^n),$$

or

$$A\alpha^n = A\alpha^{n-1} + kF(\alpha^n),$$

with α^0 given by v_h, where as before A is the mass matrix and where again F is globally Lipschitz continuous. It is clear that this nonlinear system of algebraic equations may be solved for small k, so that (4.11) defines a unique discrete solution for $nk \leq T$.

We have the following error estimate.

Theorem 4.2. Under the appropriate regularity assumptions, we have for the error in the backward Euler-Galerkin method, for small k and with $C = C(u)$,

$$\|U^n - u(nk)\| \le C(\|v_h - v\| + h^2 + k), \qquad \text{for} \quad nk \in J.$$

Proof. We write as before, with $u^n = u(nk)$,

$$(4.12) \qquad U^n - u^n = (U^n - \tilde{U}^n) + (\tilde{U}^n - u^n) = \theta^n + \rho^n,$$

where $\tilde{U}^n$ is the elliptic projection of u^n defined by

$$(4.13) \qquad (a(u^n)\nabla(\tilde{U}^n - u^n), \nabla\chi) \equiv (a(u^n)\nabla\rho^n, \nabla\chi) = 0, \quad \forall\ \chi \in S_h.$$

Note that since this is the same as (4.7), the estimates of Lemmas 4.2, 4.3, and 4.4 are still valid, and it remains bo bound θ^n. We have, for $\chi \in S_h$,

$$\begin{aligned}
&(\bar{\partial}_t\theta^n,\chi) + (a(U^n)\nabla\theta^n,\nabla\chi) \\
&= (\bar{\partial}_t U^n,\chi) + (a(U^n)\nabla U^n,\nabla\chi) - (\bar{\partial}_t\tilde{U}^n,\chi) + (a(U^n)\nabla\tilde{U}^n,\nabla\chi) \\
&= \quad (f(U^n),\chi) - (u_t^n,\chi) - (\bar{\partial}_t\tilde{U}^n - u_t^n,\chi) \\
&- (a(u^n)\nabla\tilde{U}^n,\nabla\chi) - ((a(U^n)-a(u^n))\nabla\tilde{U}^n,\nabla\chi).
\end{aligned}$$

Using (4.13) in the fourth term and the weak form of the continuous problem, we find

$$\begin{aligned}
&(\bar{\partial}_t\theta^n,\chi) + (a(U^n)\nabla\theta^n,\nabla\chi) \\
&= \quad (f(U^n)-f(u^n),\chi) - (\bar{\partial}_t(\tilde{U}^n-u^n),\chi) - (\bar{\partial}_t u^n - u_t^n,\chi) \\
&\qquad\qquad - ((a(U^n)-a(u^n))\nabla\tilde{U}^n,\nabla\chi).
\end{aligned}$$

Taking $\chi = \theta^n$ and observing the identity

$$(\bar{\partial}_t\theta^n,\theta^n) = \frac{1}{2}\,\bar{\partial}_t\|\theta^n\|^2 + \frac{k}{2}\,\|\bar{\partial}_t\theta^n\|^2,$$

this yields, by the Lipschitz continuity of f and a and the boundedness of $\nabla\tilde{U}^n$,

$$\begin{aligned}
&\frac{1}{2}\,\bar{\partial}_t\|\theta^n\|^2 + \mu\|\nabla\theta^n\|^2 \\
&\le C\|U^n-u^n\|(\|\theta^n\| + \|\rho^n\|) + (\|\bar{\partial}_t\rho^n\| + \|\bar{\partial}_t u^n - u_t^n\|)\|\theta^n\|,
\end{aligned}$$

and hence

$$\bar{\partial}_t\|\theta^n\|^2 \le C(\|\theta^n\|^2 + \|\rho^n\|^2 + \|\bar{\partial}_t\rho^n\|^2 + \|\bar{\partial}_t u^n - u_t^n\|^2)$$
$$= C(\|\theta^n\|^2 + R_n),$$

where the latter equality defines R_n. We obtain

$$(1-Ck)\|\theta^n\|^2 \le \|\theta^{n-1}\|^2 + CkR_n,$$

or, for small k,

$$\|\theta^n\|^2 \le (1+Ck)\|\theta^{n-1}\|^2 + CkR_n,$$

whence, by repeated application,

$$\|\theta^n\|^2 \le (1+Ck)^n\|\theta^0\|^2 + Ck\sum_{j=1}^{n}(1+Ck)^{n-j}R_j \tag{4.14}$$
$$\le C\|\theta^0\|^2 + Ck\sum_{j=1}^{n}R_j.$$

Here, using Lemma 4.2,

$$\|\theta^0\| = \|v_h - \tilde{U}^0\| \le \|v_h - v\| + \|\tilde{U}^0 - v\|$$
$$\le \|v_h - v\| + Ch^2\|v\|_2$$

and

$$\|\rho^j\| = \|\tilde{U}^j - u^j\| \le Ch^2\|u(jk)\|_2 \le C(u)h^2,$$

and, by Lemma 4.3,

$$\|\bar{\partial}_t\rho^j\| = \|k^{-1}\int_{(j-1)k}^{jk}\rho_t ds\| \le C(u)h^2.$$

Since finally (cf. the estimate of ω_2^j in the proof of Theorem 1.3)

$$\|\bar{\partial}_t u^n - u_t^n\| = \|k^{-1}\int_{(j-1)k}^{jk}(s-(j-1)k)u_{tt}(s)ds\| \le C(u)k,$$

we have

$$R_j \le C(u)(h^2 + k)^2,$$

and hence (4.14) shows

$$\|\theta^n\| \le C(\|v_h - v\| + h^2 + k),$$

which completes the proof.

The above method has the disadvantage that a nonlinear system of algebraic equations has to be solved at each time step, as a result of the presence of $a(U^n)$ and $f(U^n)$ in (4.11). One may therefore consider a linearized modification of the method in which this difficulty is avoided by replacing U^n by U^{n-1} in these two places, so that we now have

(4.15) $$(\bar{\partial}_t U^n,\chi) + (a(U^{n-1})\nabla U^n,\nabla\chi) = (f(U^{n-1}),\chi), \quad \forall\ \chi \in S_h.$$

With A as before and $B(\alpha)$ the positive definite matrix with elements $(a(\Sigma_\ell \alpha_\ell \phi_\ell)\nabla\phi_j,\nabla\phi_k)$, this equation may be written in the form

$$A\frac{\alpha^n - \alpha^{n-1}}{k} + B(\alpha^{n-1})\alpha^n = F(\alpha^{n-1})$$

or

$$(A + kB(\alpha^{n-1}))\alpha^n = A\alpha^{n-1} + kF(\alpha^{n-1}),$$

which may always be solved for α^n.

One may show that the result of Theorem 4.2 remains valid for this linearized form of the method (cf. Thomée [40], Chapter 10):

<u>Theorem 4.3</u>. For the linearized backward Euler-Galerkin method (4.15) we have for small k, with $C = C(u)$,

$$\|U^n - u(nk)\| \le C(\|v_h - v\| + h^2 + k).$$

For the purpose of obtaining higher accuracy in time one may consider the Crank-Nicolson-Galerkin scheme, or, with $\bar{U}^n = \frac{1}{2}(U^n+U^{n-1})$,

$$(4.16) \qquad (\bar{\partial}_t U^n,\chi) + (a(\bar{U}^n)\nabla\bar{U}^n,\nabla\chi) = (f(\bar{U}^n),\chi), \quad \forall\ \chi \in S_h, \quad n \geq 1,$$

$$U^0 = v_h.$$

This equation is symmetric around the point $t = (n-1/2)k$ and one should therefore expect second order accuracy in time. It shares, however, with the backward Euler method discussed first above, the disadvantage of producing, at each time level, a non-linear system of equations. For this reason we shall consider also a linearized modification, in which the argument of a and f is obtained by extrapolation from U^{n-1} and U^{n-2}, or, more precisely, with $\hat{U}^n = \frac{3}{2}U^{n-1} - \frac{1}{2}U^{n-2}$,

$$(4.17) \qquad (\bar{\partial}_t U^n,\chi) + (a(\hat{U}^n)\nabla\bar{U}^n,\nabla\chi) = (f(\hat{U}^n),\chi), \quad \forall\ \chi \in S_h, \quad n \geq 2.$$

As was the case for the backward Euler scheme, the nonlinear equation (4.16) will be solvable for U^n in terms of U^{n-1} for k small, whereas the linearized equation (4.17) is always solvable for U^n when U^{n-1} and U^{n-2} are given.

Note that choosing a and f at U^{n-1} as we did for the backward Euler scheme will not be satisfactory here since this would be less accurate than necessary, whereas since

$$\hat{u}^n = \frac{3}{2}u^{n-1} - \frac{1}{2}u^{n-2} = u^{n-1/2} + O(k^2), \quad \text{as} \quad k \to 0,$$

the choice just proposed will have the desired accuracy.

We observe that since the equation now contains U^{n-2}, it may only be used for $n \geq 2$, and we have to supplement it with another method for determining U^1. This may be done by a predictor corrector method, using as a first approximation the value $U^{1,0}$ determined by the case $n = 1$ of equation (4.17) with $\hat{U}^1$ replaced by U^0 and then as the final approximation the result of the same equation with $\hat{U}^1$ replaced by $\frac{1}{2}(U^{1,0} + U^0)$, so that thus our starting procedure is defined by

$$U^0 = v_h,$$

followed by

$$(4.18) \qquad k^{-1}(U^{1,0}-U^0,\chi) + (a(U^0)\nabla\tfrac{1}{2}(U^{1,0}+U^0)),\nabla\chi) = (f(U^0),\chi), \quad \forall\ \chi \in S_h,$$

and then

$$(4.19) \qquad (\bar{\partial}_t U^1,\chi) + (a(\tfrac{1}{2}(U^{1,0}+U^0))\nabla\bar{U}^1,\nabla\chi) = (f(\tfrac{1}{2}(U^{1,0}+U^0)),\chi), \quad \forall\ \chi \in S_h.$$

For this method one may show the following (cf. Thomée [40], Chapter 10):

<u>Theorem 4.4</u>. Under the appropriate regularity assumptions, we have for the present form of the linarized Crank-Nicolson-Galerkin method, for k small and with $C = C(u)$,

$$\|U^n - u(nk)\| \leq C(\|v_h - v\| + h^2 + k^2).$$

The material presented so far in this section was already essentially contained in Douglas and Dupont [12] and Wheeler [44]. For related work, see also Douglas [11], Luskin [23], Rachford [31], and Thomée and Wahlbin [41].

The methods described above, even the methods using extrapolation, have the practical disadvantage that several inner products have to be computed at each time step. For instance, in the method (4.16) most recently described, the products $(a(\hat{U}^n)\nabla U^n, \nabla\Phi_j)$ and $(f(\hat{U}^n), \Phi_j)$ have to be calculated, and even though in this extrapolation procedure $\hat{U}^n$ is a known piecewise linear function, $a(\hat{U}^n)$ and $f(\hat{U}^n)$ may be complicated. It is therefore desirable to do this by means of simple quadrature formulas.

We recall from Section 1, for the linear case, the choice $(f,\chi)_h$ for the right hand side, where

$$(\varphi,\psi)_h = \sum_{\tau\in\mathcal{T}_h} Q_{\tau,h}(\varphi\psi),$$

with $Q_{\tau,h}$ defined by the quadrature formula (1.18). Applied to the present case, this replaces the calculation of $(f(\hat{U}^n),\chi)$ by that of the expression $(f(\hat{U}^n),\chi)_h$, which depends on $f(\hat{U}^n)$ and thus $\hat{U}^n$ only at the interior vertices of the triangulation.

In the present situation it is also desirable, in order to approximate the gradient term, to have access to a similar approximation to an expression of the form $(\varphi\nabla\psi,\nabla\chi)$, where φ is a continuous function. Recalling that $\nabla\chi$ is constant on each of the triangles $\tau \in \mathcal{T}_h$ for $\chi \in S_h$, we set, with our earlier notation, for $\psi,\chi \in S_h$ and φ continuous,

$$(\varphi\nabla\psi,\nabla\chi)_h = \sum_{\tau\in\mathcal{T}_h} Q_{\tau,h}(\varphi)\nabla\psi|_\tau\cdot\nabla\chi|_\tau.$$

With this notation we may replace the semidiscrete problem (4.4) by

$$(4.20)\qquad (u_{h,t},\chi)_h + (a(u_h)\nabla u_h,\nabla\chi)_h = (f(u_h),\chi)_h, \quad \forall\ \chi \in S_h,\ \ t \in J,$$

$$u_h(0) = v_h.$$

Note that here we have also used the quadrature formula in the first term, resulting in mass lumping, so that the method (4.20) may be described as the semidiscrete lumped mass method with quadrature. We emphasize again that u_h now only enters at the interior nodes of the triangulation, so that with $\alpha = \alpha(t)$ the vector made up from these nodal values, the system (4.20) takes the form

$$\bar{A}\alpha' = F(\alpha), \qquad \text{for } t \in J,$$
$$\alpha(0) = \gamma,$$

where $\bar{A}$ is the diagonal matrix with diagonal elements $\|\phi_j\|_h^2$.

For the Crank-Nicolson method with extrapolation we have similarly

$$(4.21) \qquad (\bar{\partial}_t U^n,\chi)_h + (a(\hat{U}^n)\nabla\bar{U}^n,\nabla\chi)_h = (f(\hat{U}^n),\chi)_h, \quad \forall\, \chi \in S_h, \quad \text{for } n \geq 2,$$

with $U^0 = v_h$ and with U^1 defined by the obvious analogues of (4.18) and (4.19).

Let us remark that in this context the forward Euler method $(\partial_t U^n = (U^{n+1}-U^n)/k)$

$$(4.22) \qquad (\partial_t U^n,\chi)_h + (a(U^n)\nabla U^n,\nabla\chi)_h = (f(U^n),\chi)_h, \quad \forall\, \chi \in S_h, \quad n \geq 0.$$
$$U^0 = v_h,$$

takes the simple form

$$\bar{A}\alpha^{n+1} = F(\alpha^n), \qquad \text{for } n \geq 0,$$
$$\alpha^0 = \gamma.$$

Since $\bar{A}$ is diagonal this may be thought of as an explicit finite difference scheme on a nonuniform mesh.

For these modified methods, the same convergence results are valid as for the original methods, so that, under the appropriate regularity assumptions and with suitable choice of v_h, we have, for the semidiscrete equation (4.20),

$$\|u_h(t) - u(t)\| \le C(u)h^2, \quad \text{for } t \in J,$$

for the Crank-Nicolson method (4.21), with the appropriate starting values,

$$\|U^n - u^n\| \le C(u)(h^2 + k^2), \quad \text{for } nk \in J,$$

and finally, for the forward Euler method (4.22), under a stability condition of the form $k/h^2 \le \delta$,

$$\|U^n - u^n\| \le C(u)h^2, \quad \text{for } nk \in J.$$

For details and proofs concerning the lumped mass method with quadrature, see Nie and Thomée [25].

So far we have not made any specific statements about the regularity required from the solution, but just assumed this to be sufficient for the calculation at hand to give the optimal result. We shall now end this section with a short discussion of nonsmooth data estimates similar to those discussed for linear problems in Sections 2 and 3. In doing so we shall restrict the considerations to the semilinear equation

(4.23) $$u_t - \Delta u = f(u), \quad \text{in } \Omega, \quad 0 < t \le T.$$

The presentation will describe some recent results from Johnson, Larsson, Thomée, and Wahlbin [18] concerning the semidiscretization in space of (4.23).

We consider thus the semilinear parabolic equation (4.23) with the standard homogeneous Dirichlet boundary conditions and with $u(0) = v$. We assume that $f(y)$ is a smooth function which is bounded together with its first derivative for all $y \in R$. We consider now a family of subspaces S_h of $H_0^1(\Omega)$ which we now assume to be of order r so that the standard approximation assumption (1.2) holds. The semidiscrete solution is then defined by

$$(4.24) \qquad (u_{h,t}, \chi) + (\nabla u_h, \nabla \chi) = (f(u_h), \chi), \qquad \forall \in S_h, \quad t \in J,$$

$$u_h(0) = v_h.$$

For the case of a smooth solution the error is then, for appropriate choice of v, estimated as

$$\|u_h(t) - u(t)\| \le C(u)h^2, \qquad \text{for } t \in J.$$

For nonsmooth initial data we shall first show the following result.

Theorem 4.5. Let $v_h = P_h v$ and assume $0 \le \sigma < 2$. Then for each $\rho > 0$ there is a constant $C = C(\rho,T)$ such that

$$\|u_h(t) - u(t)\| \le Ch^\sigma t^{-\sigma/2}, \text{ if } \|v\| \le \rho, \text{ for } 0 < t \le T.$$

Proof. With E(t) the solution operator of the linear homogeneous problem the solution of (4.23), with the initial data prescribed, satisfies the integral equation

$$u(t) = E(t)v + \int_0^t E(t-s)f(u(s))ds.$$

Similarly we have, for the semidiscrete problem (4.24),

$$u_h(t) = E_h(t)P_h v + \int_0^t E_h(t-s)P_h f(u_h(s))ds.$$

Hence, setting $F_h(t) = E_h(t)P_h - E(t)$, we have for the error $e=u_h-u$,

$$e(t) = F_h(t)v + \int_0^t E_h(t-s)P_h(f(u_h(s)-f(u(s))ds + \int_0^t F_h(t-s)f(u(s))ds.$$

By our nonsmooth data error estimate of Theorem 2.1,

$$\|F_h(t)v\| \le Ch^s t^{-s/2}\|v\|, \quad \text{for } 0 \le s \le r,$$

whence we may conclude that

$$(4.25) \qquad \|e(t)\| \le Ch^\sigma t^{-\sigma/2}\|v\| + C\int_0^t \|e(s)\|ds + Ch^\sigma \int_0^t (t-s)^{-\sigma/2}ds,$$

or, since $\sigma<2$,

$$t^{\sigma/2}\|e(t)\| \le Ch^\sigma + C\int_0^t s^{-\sigma/2}(s^{\sigma/2}\|e(s)\|)ds,$$

which implies the desired result by Grönwall's lemma.

The hypothesis $\sigma < 2$ we used in the last term in (4.25) and also to secure the integrability of $s^{-\sigma/2}$ in the application of Grönwall's lemma.

The proof may easily be improved to show the somewhat stronger estimate

$$\|u_h(t) - u(t)\| \le Ch^2\{t^{-1} + |\log(h^2/t)|\}, \quad \text{if } \|v\| \le \rho.$$

In the case r=2 the error in the semidiscrete case is thus essentially of the same order as for the linear homogeneous equation. However, for r>2, this is not the case. In fact, Theorem 4.5 is essentially sharp in the sense that an estimate for the form

$$(4.26) \qquad \|u_h(t_0) - u(t_0)\| \le C(\rho,t_0)h^\sigma, \text{ for } \|v\| \le \rho,$$

cannot hold for any pair σ and t_0 with $\sigma > 2$, $t_0 > 0$.

We shall indicate this by presenting a simple example, in a somewhat modified framework. We consider the parabolic system with periodic boundary conditions, and $u = (u_1, u_2)^T$,

$$u_t - u_{xx} = \begin{bmatrix} f(u_2) \\ 0 \end{bmatrix}, \quad \text{for } |x| \leq \pi,\ t > 0,$$

$$u(-\pi,t) = u(\pi,t), \quad u_x(-\pi,t) = u_x(\pi,t),$$

$$u(0) = \begin{bmatrix} 0 \\ v_2 \end{bmatrix},$$

where $f(y) = 4y^2$ for $|y| \leq 1$, and is bounded together with its first derivative. For the approximate problem we let $h=1/n$ with n a positive integer and let S_h be the span of $\{1, \cos jx, \sin jx;\ j=1,\cdots,n-1\}$. A smooth periodic function may then be aproximated in S_h to order $O(n^{-r}) = O(h^r)$, for r arbitrary.

Letting now $v_2 = \cos nx$ we have $P_h v = 0$ and hence $u_h(x,t) \equiv 0$ for $t \geq 0$. On the other hand we find at once for the exact solution

$$u_2(x,t) = \exp(-n^2 t)\cos nx,$$

and hence

$$u_1(x,t) = n^{-2}(1 - \exp(-2n^2 t))(1 + \exp(-2n^2 t)\cos 2nx).$$

As a consequence, for any $t_0 > 0$, we have, for large n,

$$\|u_h(t_0) - u(t_0)\| = \|u_h(t_0)\| \simeq \frac{\sqrt{2\pi}}{n^2} = \sqrt{2\pi}\, h^2,$$

which contradicts (4.26) for any $\sigma > 0$.

A counterexample can also be found in [18] in the more standard setting of a scalar parabolic equation with Dirichlet boundary conditions and with piecewise polynomial approximating functions. The calculations are somewhat more lenthy.

Results similar to those above, but relating to the discretization in time, can be found in Crouzeix and Thomée [9]. In this case, $O(k)$ is the highest order of time discretization error that may be attained, in general, for initial data only in $L_2(\Omega)$.

REFERENCES.

[1] G.A. Baker, J.H. Bramble, and V. Thomée, Single step Galerkin approximations for parabolic problems. Math. Comp. 31 (1977), 818-847.

[2] J. Blair, Approximate solution of elliptic and parabolic boundary value problems, Thesis, University of California, Berkeley, 1970.

[3] J. H. Bramble and P. H. Sammon, Efficient higher order single step methods for parabolic problems: Part I. Math. Comp. 35 (1980), 665-677.

[4] J. H. Bramble, A. H. Schatz, V. Thomée and L. B. Wahlbin, Some convergence estimates for semidiscrete Galerkin type approximations for parabolic equations. SIAM J. Numer. Anal. 14 (1977), 218-241.

[5] P. Brenner, M. Crouzeix, and V. Thomée, Single step methods for inhomogeneous linear differential equations in Banach space. RAIRO Anal. Numér. 16 (1982), 5-26.

[6] C. M. Chen and V. Thomée, A lumped mass finite element method for a parabolic problem, J. Austr. Math. Soc. Ser. B. 26 (1985), 329-354.

[7] P. G. Ciarlet, The Finite Element Method for Elliptic Problems, North Holland, Amsterdam, 1978.

[8] M. Crouzeix, Sur l'approximation différentielles opérationelles linéaires par des méthodes de Runge-Kutta. Thesis. Université Paris VI, 1975.

[9] M. Crouzeix and V. Thomée, On the discretization in time of semilinear parabolic equations with nonsmooth initial data. To appear in Math. Comp.

[10] M. Dobrowolski, L^{∞}-convergence of linear finite element approximatins to quasilinear initial boundary value problems. RAIRO, Anal. Numér. 12 (1978), 247-266.

[11] J. Douglas, Jr., Effective time-stepping methods for the numerical solution of nonlinear parabolic problems. Mathematics of Finite Elements and Applications, III (Proc. Third MAFELAP Conf., Brunel Univ., Uxbridge, 1978), Academic Press, London 1979, pp. 289-304.

[12] J. Douglas, Jr. and T. Dupont, Galerkin methods for parabolic equations, SIAM J. Numer. Anal. 7 (1970), 575-626.

[13] K Eriksson, C. Johnson, and V. Thomée, A discontinuous in time Galerkin method for parabolic type problems. RAIRO, Modél. Math. Anal. Numér. 19 (1985), 611-643.

[14] G. Fix and N. Nassif, On finite element approximations in time dependent problems. Numer. Math. 19 (1972), 127-135.

[15] H. Fujii, Some remarks on finite element analysis of time-dependent field problems. Theory and Practice in Finite Element Structural Analysis, University of Tokyo Press, 1973, pp. 91-106.

[16] H.-P. Helfrich, Fehlerabschätzungen für das Galerkinverfahren zur Lösung von Evolutionsgleichungen. Manuscr. Math. 13 (1974), 219-235.

[17] M. Huang and V. Thomée, Some convergence estimates for semidiscrete type schemes for time-dependent nonselfadjoint parabolic equations. Math. Comp. 37 (1981), 327-346.

[18] C. Johnson, S. Larsson, V. Thomée and L. B. Wahlbin, Error estimates for spatially discrete approximations of semilinear parabolic equations with nonsmooth initial data.To appear in Math. Comp.

[19] C. Johnson, Y.-Y. Nie and V. Thomée, An aposterior error estimate and automatic time step control for a backward Euler discretization of a parabolic problem. Preprint 1985.

[20] M. Křížek and P. Neittaanmäki, On superconvergence techniques, Department of mathematics, University of Jyväskylä, 1984.

[21] M.-N. Le Roux, Semidiscretization in time for parabolic problems. Math. Comp. 33 (1979), 919-931.

[22] M.-N. Le Roux, Semi-discrétisation en temps pour les équations d'évolution paraboliques lorsque l'operateur dépend du temps. RAIRO, Anal. Numér. 13 (1979), 119-137.

[23] M. Luskin, A Galerkin method for nonlinear parabolic equations with nonlinear boundary conditions. SIAM J. Numer. Anal. 16 (1979), 284-299.

[24] M. Luskin and R. Rannacher, On the smoothing property of the Galerkin method for parabolic equations, SIAM J. Numer. Anal. 19 (1982), 93-113.

[25] Y.-Y. Nie and V. Thomée, A lumped mass finite-element method with quadrature for a nonlinear parabolic problem, IMA J. Numer. Anal. 5 (1985), 371-396.

[26] J. A. Nitsche, L_∞-convergence of finite element approximation, 2 Conference on Finite Elements, Rennes, France, May 12-14, 1975.

[27] J. A. Nitsche, L_∞-convergence of finite element Galerkin approximation on parabolic problems. RAIRO, Anal. Numér. 13 (1979), 31-54.

[28] J. A. Nitsche and M. W. Wheeler, L_∞-boundedness of the finite element Galerkin operator for parabolic problems. Numer. Funct. Anal. Optimization 4 (1981-82), 325-353.

[29] L. A. Oganesjan and L. A. Ruhovec, An investigation of the rate of convergence of variational-difference schemes for second order elliptic equations in a two-dimensional region with smooth boundary, Ž. Vyčisl. Mat. i Mat. Fiz. 9 (1969), 1102-1120.

[30] H. S. Price and R. S. Varga, Error bounds for semi-discrete Galerkin approximations of parabolic problems with applications to petroleum reservoir mechanics. Numerical Solution of Field Problems in Continuum Physics, AMS, Providence, R.A., 1970, 74-94.

[31] H. H. Rachford, Jr., Two-level discrete-time Galerkin approximations for second order nonlinear parabolic partial differential equations, SIAM J. Numer. Anal. 10 (1973), 1010- 1026.

[32] R. Rannacher, Finite element solution of diffusion problems with irregular data, Numer. Math. 43 (1984), 309-327.

[33] R. Rannacher and R. Scott, Some optimal error estimates for piecewise linear finite element approximations, Math. Comp. 38 (1982), 437-445.

[34] P. A. Raviart, The use of numerical integration in finite element methods for solving parabolic equations. Topics in Numerical Analysis, ed. J. J. H. Miller, Academic Press, 1973, pp. 233-264.

[35] P. H. Sammon, Convergence estimates for semidicrete parabolic equation apprximations. SIAM J. Numer. Anal. 19 (1982), 68-92.

[36] P. Sammon, Fully discrete approximation methods for parabolic problems with non-smooth initial data. SIAM J. Numer. Anal. 20 (1983), 437-470.

[37] A. H. Schatz, V. Thomée, and L. B. Wahlbin, Maximum norm stability and error estimates in parabolic finite element equations. Comm. Pure Appl. Math. 33 (1980), 265-304.

[38] V. Thomée, Some convergence results for Galerkin methods for parabolic boundary value problems. Mathematical Aspects of Finite Elements in Partial Differential Equations, C. de Boor ed., Academic Press, 1974, pp. 55-88.

[39] V. Thomée, Negative norm estimates and superconvergence in Galerkin methods for parabolic problems. Math. Comp. 34 (1980), 93-113.

[40] V. Thomée, Galerkin Finite Element Methods for Parabolic Problems, Springer Lecture Notes in Mathematics No. 1054, 1984.

[41] V. Thomée and L. B. Wahlbin, On Galerkin methods in semilinear parabolic problems. SIAM J. Numer. Anal. 12 (1975), 378-389.

[42] V. Thomée and L. B. Wahlbin, Maximum-norm stability and error estimates in Galerkin methods for parabolic equations in one space variable. Numer. Math. 41 (1983), 345-371.

[43] L. B. Wahlbin, A remark on parabolic smoothing and the finite element method, SIAM J. Numer. Anal. 17 (1980), 378-389.

[44] M. F. Wheeler, A priori L_2 error estimates for Galerkin approximations to parabolic partial differential equations, SIAM J. Numer. Anal. 10 (1973), 723-759.

[45] M. H. Wheeler, L_∞ estimates of optimal orders for Galerkin methods for one dimensional second order parabolic and hyperbolic equations. SIAM J. Numer. Anal. 10 (1973), 908-913.

[46] M. Zlámal, Finite element multistep discretization of parabolic boundary value problems. Math. Comp. 29 (1975), 350-359.

Vidar Thomée
Department of Mathematics
Chalmers University of Technology
S-41296 Göteborg, Sweden

Boundary Element Methods for Elliptic Problems

W. L. Wendland
University of Stuttgart
Mathematisches Institut A
Pfaffenwaldring 57
D-7000 Stuttgart 80
Federal Republic of Germany

Introduction

During the last decades, the numerical treatment of boundary integral equations in the form of boundary element methods has become a rather popular and powerful technique for engineering computations of boundary value problems, in addition to finite difference and finite element methods. In these lecture notes are presented some of the basic mathematical concepts. We begin with several elliptic boundary value problems of classical mathematical physics and their reduction to the boundary. For simplicity, we mostly consider only two-dimensional problems since there are no significant difficulties in formulating the corresponding three-dimensional problems and boundary integral equations. All these integral equations as some Fredholm integral equations of the first and the second kind, certain Cauchy singular integral equations and equations with hypersingular kernels belong to the class of classical pseudo-differential operators on the boundary which provide a comprehensive formulation and characterization. The concept of the principal symbol allows the characterization of boundary integral equations whose variational formulation on the boundary provides there a Gårding inequality. Therefore Ritz-Galerkin boundary element methods can be analyzed similarly to the domain finite element methods providing asymptotic convergence if the number of grid points increases. In engineering codes, however, mostly point collocation is used for the boundary element methods. Here stability is much more difficult to prove. In two-dimensional problems where Γ is a curve, we now have two different techniques available. One is based on the equivalent formulation with a modified Ritz-Galerkin method exploiting that derivatives of smoothest polynomial splines become Dirac functionals at the knots, and the other is based on the explicit analysis of the Fourier coefficients of the Fourier transformed collocation equations. Since the latter requires tedious technicalities in Fourier analysis we here only present a short sketch of the stability proof.

Since for the boundary element methods so many different branches of analysis need to be combined, these notes can only provide a brief introduction and some of the details are left out. However, the asymptotic error analysis based on the Ritz-Galerkin formulation for strongly elliptic boundary integral equations is presented in detail.

Engineering aspects of the boundary element methods can be found e.g. in the books

[7], [35], [48]. The aim of these notes, however, is the mathematical foundation, part of which can also be found in [62] and for numerical analysis in [31].

I hope that the student will catch some of the fascination of this recent field in numerical analysis.

I would like to thank the participants of the seminar for their encouraging and stimulating discussions and I want to thank Ms. O. Steidel and Mr. J. Witt for typing and Dr. W. Kolbe for proofreading this manuscript.

1 Boundary Integral Equations

For the sake of simplicity, we present here only a few different problems of classical mathematical physics which can be treated by boundary integral equations leading to boundary element methods. Moreover, in this chapter we consider only two-dimensional problems.

1.1 The exterior Neumann problem for the Laplacian

Let us consider an exterior stationary divergence-free irrotational two-dimensional flow in the domain Ω^c exterior to a given obstacle – or profile – Ω with boundary Γ. Let us denote the desired velocity field by $\vec{v}(\vec{x}) = (v_1, v_2)^T$ where ^{T} will denote the transposed vector (or matrix); and where $\vec{x} = (x_1, x_2)^T \in \mathbb{R}^2$ denotes the position vector; we choose the origin $\vec{0} \in \Omega$. The velocity field $\vec{v}$ has a potential and can be expressed by

(1.1.1) $$\vec{v}(\vec{x}) = v_\infty \vec{i} + \frac{\lambda}{2\pi} \frac{1}{|\vec{x}|^2} \begin{pmatrix} -x_2 \\ x_1 \end{pmatrix} + \nabla u \quad \text{in } \Omega^c$$

where v_∞ is the given velocity at infinity in the direction of the unity-vector $\vec{i}$ in x_1-direction. $\nabla = (\partial_{x_1}, \partial_{x_2})^T$ denotes the nabla-operator. Here λ is some given circulation which, more generally, will often be evaluated in an additional step with the help of the so-called Kutta-Joukowski condition at the trailing edge of the given profile Γ.

The potential $u(\vec{x})$ in (1.1.1) is the desired solution of the exterior Neumann problem:

(1.1.2) $$\Delta u = 0 \qquad \text{in } \Omega^c,$$

(1.1.3) $$\partial_n u|_\Gamma := \vec{n}(\vec{x}) \cdot \nabla u(\vec{x})_{|\vec{x}\in\Gamma} = g(\vec{x}), \quad \vec{x} \in \Gamma$$

where

$$g(\vec{x}) = -v_\infty n_1(\vec{x}) - \frac{\lambda}{2\pi} \frac{1}{|\vec{x}|^2} (x_1 n_2(\vec{x}) - x_2 n_1(\vec{x}))$$

is given on Γ and where u satisfies at infinity the decay condition

(1.1.4) $$u(\vec{x}) = O(|\vec{x}|^{-1}) \text{ as } |\vec{x}| \to \infty.$$

$\vec{n}(\vec{x}) = (n_1(\vec{x}), n_2(\vec{x}))^T$ denotes the unit-normal vector on Γ exterior to Ω, and Γ will in the following be assumed to be a sufficiently smooth Jordan curve whose properties are to be specified in the sequel. Δ denotes the Laplacian.

The classical linear boundary value problem (1.1.2)-(1.1.4) can be reformulated as a boundary integral equation with the aid of the Green representation theorem, which requires some minimal smoothness of Γ and the knowledge of the fundamental solution of the Laplacian in (1.1.2).

Assumptions for Γ:
Let Γ be a Jordan curve having the L-periodic parametric representation with respect to the arc-length s,

(1.1.5) $$\vec{x} = \vec{\xi}(s) \quad \text{with } \vec{\xi} \in C^{1+\alpha},$$

i.e. $\vec{\xi} = (\xi_1, \xi_2)^T$ *is Hölder-continuously differentiable with an Hölder exponent* $\alpha \in (0,1)$.
L denotes the length of Γ.

Lemma 1.1 ([61]) *If Γ satisfies* (1.1.5) *then Γ has the Lyapounov properties,*

(1.1.6)
(i) $\vec{n}(s) = \left(\dot{\xi}_2(s), -\dot{\xi}_1(s)\right)^T \in C^\alpha$;

(ii) $\left|\vec{n}(s) \cdot \left(\vec{\xi}(s) - \vec{\xi}(\sigma)\right)\right| \le c\left|\vec{\xi}(s) - \vec{\xi}(\sigma)\right|^{1+\alpha}$ *with some global constant* c;

(iii) *there exists a positive constant $d_0(\Gamma)$ such that for every d with $0 < d \le d_0(\Gamma)$, and every σ, the equation $|\vec{\xi}(\sigma) - \vec{\xi}(s)| = d$ has exactly two points of solution $\vec{\xi}(s_1) \neq \vec{\xi}(s_2)$ on Γ.*

For the proof see [61].

Definition of the fundamental solution:
Let $\delta_{\vec{y}}(\vec{x})$ denote the Dirac distribution concentrated at $\vec{y} \in \mathbb{R}^2$. Then the solution F of

(1.1.7) $$-\Delta_{\vec{x}} F = \delta_{\vec{y}}(\vec{x})$$

is called the *fundamental solution* (see also [63] p. 142ff.).

Of course, the fundamental solution is not uniquely determined by (1.1.7); in the sequel we can choose any of them. However, for partial differential operators with constant coefficients it is advantageous for many purposes to choose the translation-invariant fundamental solution.

For the Laplacian in $\mathbb{R}^2$, i.e. for (1.1.7), the latter is given by

$$F(\vec{x},\vec{y}) = -\frac{1}{2\pi}\log|\vec{x}-\vec{y}|. \tag{1.1.8}$$

Theorem 1.2 (The Green representation theorem) *Let $u \in C^1(\Omega^c \cup \Gamma) \cap C^2(\Omega^c)$ be a solution of* (1.1.2) *in Ω^c with $u(\vec{x}) = O(|\vec{x}|^{-1})$ for $|\vec{x}| \to \infty$, and let Γ satisfy the assumptions* (1.1.5). *Then u admits the representation*

$$\begin{aligned} u(\vec{x}) &= -\frac{1}{2\pi}\int_{\vec{y}\in\Gamma} u(\vec{y})\left(\partial_{n_y}\log|\vec{x}-\vec{y}|\right) ds_{\vec{y}} \\ &\quad +\frac{1}{2\pi}\int_{\vec{y}\in\Gamma}\log|\vec{x}-\vec{y}|\,(\partial_n u(\vec{y}))\, ds_{\vec{y}} \quad \text{for all } \vec{x}\in\Omega^c. \end{aligned} \tag{1.1.9}$$

This is a classical result of potential theory. For the proof see e.g. [61] or [19] p. 257.

As we can see from (1.1.9), the solution u of (1.1.2) is known everywhere on Ω^c and can be represented by the double layer potential and the single layer potential on the right-hand side of (1.1.9) if the *Cauchy data* $(u,\partial_n u)^T_{|\Gamma}$ are known on Γ.

The assumptions for Γ and for the solution can be relaxed significantly; (1.1.9) is still valid for a Lipschitz boundary Γ and for generalized solutions u (see e.g. [11]).

For our specific Neumann problem (1.1.2)-(1.1.4), one Cauchy datum $\partial_n u|_\Gamma = g$ is given by (1.1.3), whereas the other Cauchy datum, $u|_\Gamma$ is yet unknown. For its determination we can obtain boundary integral equations from (1.1.9) by taking $\vec{x} \to \Gamma$ – or by taking the normal derivative ∂_n of both sides of (1.1.9) on Γ. For this purpose we need the corresponding *boundary traces* of the potentials of the right-hand side of (1.1.9) which are known in terms of the so-called 'jump relations'. For the Laplacian in (1.1.2), these can be obtained by using a more geometrical interpretation of the double layer kernel $(\partial_n \log|\vec{x}-\vec{y}|)$ in terms of polar coordinates about $\vec{x}$:

$$\vec{y} = \vec{x} + (\rho\cos\Theta_{\vec{x}}, \rho\sin\Theta_{\vec{x}})^T$$

with $\rho = |\vec{x}-\vec{y}|$ and $\Theta_{\vec{x}}$ the angular coordinate. Then elementary transformation yields for $\vec{y} = \vec{\xi}(s)$ the relation

$$\partial_{n_y}(\log|\vec{x}-\vec{y}|)_{|\Gamma} = \frac{d}{ds}\Theta_{\vec{x}}(\vec{y})_{|\Gamma} := \frac{\partial\Theta_{\vec{x}}}{\partial y_1}\dot{\xi}_1(s) + \frac{\partial\Theta_{\vec{x}}}{\partial y_2}\dot{\xi}_2(s). \tag{1.1.10}$$

This is the angle under which the boundary curve element $ds_{\vec{y}}$ at $\vec{y}\in\Gamma$ can be visualized when looking at $ds_{\vec{y}}$ from $\vec{x}\notin\Gamma$; the sign is given by the sign of the scalar product $(\vec{y}-\vec{x})\cdot\vec{n}(\vec{y})$.

Then the basic property for the validity of the jump relations is boundedness of the total variation of $\Theta_{\vec{x}}\left(\vec{\xi}(s)\right)$ uniformly for all $\vec{x}$. For Γ satisfying (1.1.5) we have:

Lemma 1.3 *For the Lyapounov curve* Γ, *i.e. with* (1.1.5), *we have*

$$A := \sup_{\vec{x} \notin \Gamma} \int_\Gamma \left| d\Theta_{\vec{x}} \left(\vec{\xi}(s)\right) \right| < \infty. \tag{1.1.11}$$

In addition, for convex Ω *we have*

$$A = 2\pi = \int_\Gamma d\Theta_{\vec{x}} \left(\vec{\xi}(s)\right) \quad \text{for all } \vec{x} \in \Omega. \tag{1.1.12}$$

For the proof see [61].

Remark: It is an easy consequence that (1.1.11) also holds for piecewise Lyapounov curves. J. Kral gave in [52] necessary and sufficient conditions for Γ in order (1.1.11) to hold. I. Netuka showed in [66] that (1.1.11), (1.1.12) are valid for *any* convex Ω.

The following relation is due to C. F. Gauss.

Proposition 1.4 *Let* Γ *satisfy* (1.1.5). *Then*

$$\tilde{\Theta}_{\vec{x}} := \int_{\Gamma \setminus \{\vec{x}\}} d\Theta_{\vec{x}} \left(\vec{\xi}(s)\right) = \begin{cases} 2\pi & \text{for } \vec{x} \in \Omega, \\ \pi & \text{for } \vec{x} \in \Gamma, \\ 0 & \text{for } \vec{x} \in \Omega^c. \end{cases} \tag{1.1.13}$$

The first and the last relation follow immediately from (1.1.9) with $u \equiv 1$; for $\vec{x} \in \Omega^c$ consider a ring domain with $\vec{y} \in \Omega^c \wedge |\vec{y} - \vec{x}| \le R_0$ instead of the whole exterior Ω^c, where R_0 is chosen sufficiently large. For $\vec{x} \in \Gamma$ use appropriate bounds for $\Theta_{\vec{x}}$.

Remark: For non-smooth Γ and $\vec{x}_0 \in \Gamma$, the integral in (1.1.13) might still exist, but $\tilde{\Theta}_{\vec{x}_0} \neq \pi$, if $\vec{x}_0$ is a corner point. This is the case for curves of bounded rotation as was shown by J. Radon [74], and for the curves satisfying (1.1.11) and characterized in [52]; containing in particular piecewise Lyapounov curves.

Now we are in the position to execute the limits $\vec{x} \to \vec{x}_0 \in \Gamma$ for $\vec{x} \in \Omega$ and for $\vec{x} \in \Omega^c$, respectively.

Theorem 1.5 (Jump relations) *Let* $\mu \in C^0(\Gamma)$ *be given, and let* Γ *be piecewise Lyapounov. Then the double layer potential* $\int_\Gamma \mu(s)\, d\Theta_{\vec{x}} \left(\vec{\xi}(s)\right)$ *can be extended continuously up to the boundary* Γ *from* $\vec{x} \in \Omega$ *and from* $\vec{x} \in \Omega^c$, *respectively. The respective boundary values satisfy the jump relation*

$$\begin{aligned} &\lim_{\vec{x} \to \vec{x}_0} \int_\Gamma \mu(s)\, d\Theta_{\vec{x}} \left(\vec{\xi}(s)\right) = \\ &\qquad \pm\pi\mu(\vec{x}_0) + \int_{\Gamma \setminus \{\vec{x}_0\}} \mu(s)\, d\Theta_{\vec{x}_0} \left(\vec{\xi}(s)\right) + (\pi - \tilde{\Theta}_{\vec{x}_0})\mu(\vec{x}_0) \end{aligned} \tag{1.1.14}$$

for $\vec{x}_0 \in \Gamma$ *and for* $\vec{x} \in \Omega$ *with the* + sign; *and for* $\vec{x} \in \Omega^c$ *with the* − sign, *respectively.*

For the proof see e.g. [61]. Note, that for $\mu(\vec{x}_0) \neq 0$ the limits from Ω and from Ω^c are different. Moreover, at smooth points $\vec{x}_0$ of Γ, the last term vanishes.

Now we take $\Omega^c \ni \vec{x} \to \vec{x}_0$ at both sides of (1.1.9).

Lemma 1.6 *Let $u \in C^1(\Omega^c \cup \Gamma) \cap C^2(\Omega^c)$ be a solution of* (1.1.2)-(1.1.4). *Then the Cauchy data $(u,g)_{|\Gamma}$ satisfy the boundary integral equation*

$$(1.1.15)\qquad \begin{aligned} u(\vec{x}_0) &= -\frac{1}{\pi}\int_{\Gamma\setminus\{\vec{x}_0\}} u\left(\vec{\xi}(s)\right)\, d\Theta_{\vec{x}_0}\left(\vec{\xi}(s)\right) - \frac{1}{\pi}(\pi - \tilde{\Theta}_{\vec{x}_0})u(\vec{x}_0) \\ &\quad +\frac{1}{\pi}\lim_{\vec{x}\to\vec{x}_0}\int_\Gamma \log|\vec{x}-\vec{y}|g(\vec{y})ds_{\vec{y}}. \end{aligned}$$

The last limit on the right-hand side exists, since all the other limits in (1.1.9) exist. In fact, this single layer potential is continuous across Γ for a large class of Γ and g.

Definition: A Jordan curve Γ is called a *c-curve* if there exist positive constants $c_1, c_2 > 0$ such that

$$(1.1.16)\qquad c_1|s-\sigma| \leq \left|\vec{\xi}(s) - \vec{\xi}(\sigma)\right| \leq c_2|s-\sigma|,$$

where $s - \sigma$ is evaluated mod L.
The Hölder continuity of the tangent vector ensures that every simple closed Lyapounov curve Γ is a c-curve. For a piecewise Lyapounov boundary Γ, however, (1.1.16) excludes $\tilde{\Theta}_{\vec{x}_0} = 0$ or 2π, i.e. spines. For c-curves we have the following result for the single layer potential:

Lemma 1.7 (see D. Gaier [26]) *If Γ is a c-curve and $\varphi \in L^p(\Gamma)$ with $1 < p$, then the single layer potential*

$$(1.1.17)\qquad V\varphi(\vec{x}) := -\frac{1}{\pi}\int_\Gamma \varphi(\vec{y})\log|\vec{x}-\vec{y}|\, ds_{\vec{y}}$$

is continuous in $\mathbb{R}^2$; i.e. in particular across Γ.

For the proof see [26].

Collecting the above results we have:

Lemma 1.8 *Let Γ be a piecewise Lyapounov simple closed Jordan c-curve. Let $u \in C^1(\Omega^c \cup \Gamma) \cap C^2(\Omega^c)$ be a solution of* (1.1.2)-(1.1.4). *Then the boundary values of the potential $u_{|\Gamma}$ satisfy the boundary integral equation*

$$(1.1.18)\qquad u(\vec{x}_0) = Ku(\vec{x}_0) - Vg(\vec{x}_0) \ \text{ for all } \ \vec{x}_0 \in \Gamma.$$

Here, the double layer boundary integral operator K is given by

$$K\mu(\vec{x}_0) = -\frac{1}{\pi}\int_{\Gamma\setminus\{x_0\}} \mu\left(\vec{\xi}(s)\right)\, d\Theta_{\vec{x}_0}\left(\vec{\xi}(s)\right) - \frac{1}{\pi}\left(\pi - \tilde{\Theta}_{\vec{x}_0}\right)\mu(\vec{x}_0). \tag{1.1.19}$$

Since g is given, (1.1.18) *can serve as an integral equation for the determination of the yet unknown Cauchy datum* $u_{|\Gamma}$. Once (1.1.18) is solved for $u_{|\Gamma}$, the potential $u(\vec{x})$ can be determined for all $\vec{x} \in \Omega^c$ by (1.1.9), and its gradient is the desired velocity field due to (1.1.1). In fact, (1.1.18) is the classical Fredholm integral equation of the second kind of potential theory; for its long and attractive history see [20] and [51]. In case of a closed Lyapounov curve Γ without corners, i.e. with (1.1.15), the kernel of K in (1.1.19) defined by

$$k(\vec{x}_0, \vec{y}) = \frac{d}{ds}\Theta_{\vec{x}_0}(\vec{\xi}(s)) = -\frac{1}{\pi}\frac{\vec{n}(\vec{y})\cdot(\vec{y}-\vec{x}_0)}{|\vec{y}-\vec{x}_0|^2} \quad \text{with } \vec{y} = \vec{\xi}(s) \tag{1.1.20}$$

turns out to be a so-called *weakly singular kernel* (see [30], [61]) defining a completely continuous operator $K : C^0(\Gamma) \to C^0(\Gamma)$. If Γ is C^2, then $k(\vec{x}_0, \vec{y})$ is even a continuous function of both points, $(\vec{x}_0, \vec{y}) \in \Gamma \times \Gamma$; if $\Gamma \in C^\infty$, then $k(\vec{x}_0, \vec{y})$ is C^∞ on $\Gamma \times \Gamma$. In these cases, the boundary integral equation (1.1.18) is a classical Fredholm integral equation of the second kind with a completely continuous integral operator K in $C^0(\Gamma)$. In addition, the solution of (1.1.18) is unique (see e.g. [53] Theorem 6.16). Hence, for any given g within continuous $Vg_{|\Gamma}$, the boundary integral equation (1.1.18) admits exactly one continuous solution $u_{|\Gamma}$ which is just the second Cauchy datum for the desired potential $u(\vec{x})$ given by (1.1.9).

Note that this approach remains valid also for non-smooth Γ [52]. However, our solution procedure so far is not complete: the continuous boundary trace $u_{|\Gamma}$ generates via (1.1.1) a double layer potential in $C^0(\Omega^c \cup \Gamma) \cap C^2(\Omega^c)$ whereas Theorem 1.2 requires $C^1(\Omega^c \cup \Gamma) \cap C^2(\Omega^c)$. This gap in regularity can be closed by using weak variational solutions in appropriate Sobolev spaces; here we require $u \cdot (1 + |\vec{x}|^2)^{-2} \in H^1(\Omega^c)$, $u_{|\Gamma} \in H^{\frac{1}{2}}(\Gamma)$ and $g \in H^{-\frac{1}{2}}(\Gamma)$. In these spaces, the weak variational solution of the boundary value problem (1.1.2), (1.1.3) and the solution of the bouundary integral equation (1.1.18) will be equivalent. (For details see [28] and [13].).

Note that the direct exploitation of the Green representation formula (1.1.19) provides a second oppurtunity of a boundary integral equation by taking the normal derivative ∂_{n_x} of both sides. This yields the boundary integral equation of the first kind for the Cauchy datum $u_{|\Gamma}$,

$$\begin{aligned} Du(\vec{x}) &:= \frac{1}{\pi}\partial_{n_x}\int_{\Gamma\setminus\{\vec{x}\}} u(\vec{y})\left(\partial_{n_y}\log|\vec{x}-\vec{y}|\right) ds_{\vec{y}} \\ &= -g(\vec{x}) - K'g(\vec{x}),\ \vec{x} \in \Gamma, \end{aligned} \tag{1.1.21}$$

which is defined by the hypersingular boundary integral operator

$$(1.1.22)\quad Du(\vec{x}) = -\frac{1}{\pi} p.v. \int_{\Gamma\setminus\{\vec{x}\}} \left\{ \frac{\vec{n}(\vec{x})\cdot\vec{n}(\vec{y})}{|\vec{x}-\vec{y}|^2} + 4k(\vec{x},\vec{y})k(\vec{y},\vec{x}) \right\} \cdot \{u(\vec{y}) - u(\vec{x})\}\, ds_{\vec{y}}$$

where the right-hand side in (1.1.22) is defined as a Cauchy principal value integral. The operator K' on the right-hand side of (1.1.21) is the operator adjoint to the double layer potential operator K.

Lemma 1.9 *Let Γ be a piecewise Lyapounov and simple closed Jordan c-curve. Let $u \in C^1(\Omega^c \cup \Gamma) \cap C^2(\Omega^c)$ be a solution of* (1.1.2)-(1.1.4). *Then the trace $u_{|\Gamma}$ satisfies the boundary integral equation* (1.1.21) *at all smooth points of* Γ.

The proof follows from taking ∂_{n_x} of both sides of (1.1.9), since $\partial_{n_x} u_{|\Gamma} = g$ exists due to our assumptions and since for $g \in C^0(\Gamma)$

$$\partial_{n_x} \frac{1}{\pi} \int_{\Gamma\setminus\{\vec{x}\}} \log|\vec{x}-\vec{y}| g(\vec{y})\, ds_{\vec{y}} = g(\vec{x}) - K'\vec{g}(\vec{x})$$

holds at the smooth points $\vec{x}$ of Γ (see [51]).

The equation (1.1.21) does not belong to the standard types of boundary integral equations, but it will be one of those feasible for the boundary element treatment. We also observe that even the so-called *direct approach* based on the Green representation (1.1.9) yields more than one boundary integral equation for the missing Cauchy datum. Correspondingly, the boundary element approximation will also yield *different numerical procedures* according to the choice of the underlying boundary integral equation.

Again, different boundary integral equations will be obtained with the direct approach if one uses the *stream function* $\psi(\vec{x})$ for the velocity field;

$$(1.1.23)\qquad \vec{v}(\vec{x}) = v_\infty \vec{i} + \frac{\lambda}{2\pi}\frac{1}{|\vec{x}|^2}\begin{pmatrix} -x_2 \\ x_1 \end{pmatrix} + \psi_{x_2}\vec{i} - \psi_{x_1}\vec{j}.$$

Then the desired stream function ψ is to be determined as the solution of the exterior Dirichlet problem:

$$(1.1.24)\qquad \Delta\psi = 0 \ \text{ in } \Omega^c,$$

$$(1.1.25)\qquad \psi_{|\Gamma} = \frac{1}{2}\omega + g := \frac{1}{2}\omega - v_\infty x_2 + \frac{\lambda}{2\pi}\log|\vec{x}| \quad \text{for } \vec{x} \in \Gamma,$$

where ω is some unknown constant, where g is given, and where ψ satisfies at infinity the decay condition

$$(1.1.26)\qquad \psi(\vec{x}) = O\left(|\vec{x}|^{-1}\right) \ \text{ as } |\vec{x}| \to \infty.$$

Now Theorem 1.2 for ψ in Ω^c in connection with (1.1.13) yields the representation

$$\begin{aligned}\psi(\vec{x}) &= -\frac{1}{2\pi}\int_{\vec{y}\in\Gamma} g(\vec{y})\,d\Theta_{\vec{x}}(\vec{y}) \\ &\quad+\frac{1}{2\pi}\int_{\vec{y}\in\Gamma}(\partial_n\psi(\vec{y}))\log|\vec{x}-\vec{y}|\,ds_{\vec{y}} \quad\text{for } \vec{x}\in\Omega^c.\end{aligned} \tag{1.1.27}$$

Since here g is given, the unknown Cauchy datum in this case is the yet unknown normal derviative of ψ on Γ, namely

$$u := \partial_n\psi_{|\Gamma}. \tag{1.1.28}$$

The application of Theorem 1.5 to (1.1.27) yields:
If Γ is a piecewise Lyapounov simple closed Jordan c-curve and if $\psi \in C^1(\Omega^c \cup \Gamma)\cap C^2(\Omega^c)$ is a solution of (1.1.24)-(1.1.26), then the Cauchy data $(g,u)_{|\Gamma}$ of ψ satisfy the boundary integral equation

$$Vu(\vec{x}) + \omega = -g(\vec{x}) + Kg(\vec{x}) \quad\text{for } \vec{x}\in\Gamma. \tag{1.1.29}$$

In addition, the condition (1.1.26) yields for ψ from (1.1.27) the equilibrium condition

$$\int_\Gamma \partial_n\psi ds = \int_\Gamma u\,ds = 0. \tag{1.1.30}$$

Hence, for $u = \partial_n\psi_{|\Gamma}$ and the constant ω we arrive at a Fredholm boundary integral equation of the first kind (1.1.29) with the logarithmic kernel together with the additional equation (1.1.30). (For these equations see [41] and [43].)

If we take $\partial_{n_{\vec{x}}}|_\Gamma$ of both sides in (1.1.27) we find the Fredholm integral equation of the second kind for u,

$$u(\vec{x}) + K'u(\vec{x}) = -Dg(\vec{x}) \quad\text{for } \vec{x}\in\Gamma, \tag{1.1.31}$$

together with (1.1.30). Again, we may solve (1.1.29), (1.1.30) for u as well as (1.1.31), (1.1.30) – we have the choice.

On curves with corners, however, (1.1.31) can only be understood in a generalized sense [51] and, in particular, K' is then no longer a continuous operator in $C^0(\Gamma)$ which restricts the practical use of (1.1.31). On the other hand, if the circulation λ in (1.1.25) is considered to be an *additional* unknown and $u(TE) = 0$ – which corresponds to the Kutta-Joukowski condition at the trailing edge TE – is incorporated into the solution procedure, then (1.1.31) can still be used (see [58]).

1.2 Exterior viscous flow problems

In viscous flow problems, the boundary element method provides an efficient tool for solving Stokes problems. As an example, let us consider here the Stokes expansion of two-dimensional exterior viscous flows around a given obstacle with boundary Γ governed by the Navier-Stokes equations

$$
\begin{aligned}
&(1.2.1) && \Delta \vec{v} - \nabla p &&= Re(\vec{v}\cdot\nabla)\vec{v} \text{ and}\\
&(1.2.2) && \nabla\cdot\vec{v} &&= 0 \text{ in } \Omega^c,\\
&(1.2.3) && \vec{v}_{|\Gamma} &&= \vec{0} \text{ and } \vec{v}(\vec{x}) - \vec{i} = o(1), \quad p(\vec{x}) = o(1) \text{ as } |\vec{x}| \to \infty.
\end{aligned}
$$

Here $\vec{v}(\vec{x})$ is the desired dimension-less velocity field, Re denotes the Reynolds number which is supposed to be small, and $v_\infty = 1$. As was shown by Hsiao and MacCamy in [41], [42], the solution of (1.2.1)-(1.2.3) can be approximated by the asymptotic Stokes expansion

$$(1.2.4) \qquad \vec{v}(\vec{x}) \sim \sum_{n\geq 1} \vec{v}_n(\vec{x})(\log Re)^{-n}, p(\vec{x}) \sim \sum_{n\geq 1} p_n(\vec{x})(\log Re)^{-n},$$

where the terms $(\vec{v}_n(\vec{x}), p_n(\vec{x}))$ are solutions of the following sequence of Stokes problems:

$$(1.2.5) \qquad \begin{aligned} &\Delta\vec{v}_n - \nabla p_n = \vec{0}, \ \nabla\cdot\vec{v}_n = 0 \text{ in } \Omega^c,\\ &\vec{v}_n|_\Gamma = \vec{0} \text{ and } \vec{v}_n = \vec{A}_n \log|\vec{x}| + O(1), \ p_n(\vec{x}) = o(1) \text{ as } |\vec{x}| \to \infty. \end{aligned}$$

The vectors $\vec{A}_n$ are given consecutively by

$$\vec{A}_1 = -(1,0)^T$$

and the following boundary integral equations for corresponding stream functions ψ_n,

$$(1.2.6) \qquad \vec{v}_n = (\psi_{nx_2}, -\psi_{nx_1}).$$

Then

$$\Delta^2\psi_n = 0 \text{ in } \Omega^c;$$

and we have the representation

$$(1.2.7) \qquad \begin{aligned} \psi_n(\vec{x}) &= \int_\Gamma \nabla_{\vec{\xi}} F(\vec{x},\vec{\xi})\vec{g}_n(\vec{\xi})ds_{\vec{\xi}} - \vec{\omega}_n\cdot\vec{x} \text{ for } \vec{x}\in\Omega^c,\\ p_n(\vec{x}) &= -\frac{1}{2\pi}\int_\Gamma \vec{g}_n(\vec{\xi})\cdot\frac{(\vec{x}-\vec{\xi})}{|\vec{x}-\vec{\xi}|^2}ds_{\vec{\xi}}, \end{aligned}$$

where

$$(1.2.8) \qquad F(\vec{x},\vec{\xi}) = -\frac{1}{2}|\vec{x}-\vec{\xi}|^2\log|\vec{x}-\vec{\xi}| - \frac{1}{2}\left(\gamma - \log 4 - \frac{3}{2}\right)|\vec{x}-\vec{\xi}|^2$$

is a fundamental solution for the bi-Laplacian, where γ is Euler's constant, and where

$$\vec{g}(\vec{\xi}) \sim \sum_{n\geq 1} \vec{g}_n(\vec{\xi})(\log Re)^{-n}$$

corresponds to the asymptotic expansion of the hydrodynamic stress $\vec{\tau}(\vec{\xi}) = \left(g_2(\vec{\xi}), -g_1(\vec{\xi})\right)^T$ on the boundary Γ. The governing boundary integral equations read as

$$
(1.2.9) \qquad \begin{aligned}
&\int_\Gamma \log|\vec{x}-\vec{\xi}|\vec{g}_n(\vec{\xi})\,ds_{\vec{\xi}} + \int_\Gamma L(\vec{x},\vec{\xi})\vec{g}_n(\vec{\xi})\,ds_{\vec{\xi}} - \vec{\omega}_n = \vec{0} \text{ for } \vec{x}\in\Gamma \text{ and}\\
&\int_\Gamma \vec{g}_n\,ds = (A_{n2}, -A_{n1})^T, \int_\Gamma \vec{g}_n\cdot d\vec{\xi} = 0.
\end{aligned}
$$

For $n=1$, $\vec{A}_1 = -(1,0)^T$ and for $n=2$, $\vec{A}_2 = (\omega_{12}, -\omega_{11})^T$. The additional kernel is given by the matrix-valued function

$$
L_{\alpha\beta}(\vec{x},\vec{\xi}) = \frac{(x_\alpha - \xi_\alpha)(x_\beta - \xi_\beta)}{|\vec{x}-\vec{\xi}|^2} + \frac{1}{2}\delta_{\alpha\beta},
$$

which is smooth for Γ sufficiently smooth. $\delta_{\alpha\beta}$ denotes the Kronecker symbol. Hsiao and MacCamy obtain the sequence (1.2.9) from a singular perturbation analysis of (1.2.1)-(1.2.3) and matching with an appropriate Oseen expansion at infinity. In particular, they show in [42] the following rigorous result.

Lemma 1.10 *Let Γ be smooth, e.g. C^∞. Then the first two terms of the Stokes expansion* (1.2.4) *satisfy*

$$
(1.2.10) \qquad \left|\vec{v}(\vec{x}) - \sum_{n=1}^{2}\vec{v}_n(\vec{x})(\log Re)^{-n}\right| = O\left(|\log Re|^{-3}\right)
$$

on every compact subregion of $\overline{\Omega^c} = \Omega^c\cup\Gamma$. The hydrodynamic force is given by

$$
(1.2.11) \qquad \vec{F} = 4\pi\left\{-\vec{i}(\log Re)^{-1} + \vec{A}_1(\log Re)^{-2}\right\} + O\left(|\log Re|^{-3}\right).
$$

The boundary element treatment of the Stokes expansion (1.2.4) can be found in [40].

1.3 Scattering problems in acoustics

The scattering of an incoming plane wave $e^{i\vec{k}\cdot\vec{x}}$ at a soft or hard scatterer with boundary Γ is one of the classical standard problems in acoustics. Here we will only formulate some corresponding boundary value problems and boundary integral equations. For further details we refer to the articles in [24], to [14], [29] and to the book by Colton and Kress [10].

Harmonic acoustic vibrations are governed by the Helmholtz equation

$$
(1.3.1) \qquad (\Delta + k^2)U = 0 \text{ in } \Omega^c
$$

for the acoustic pressure amplitude U where $k \geq 0$ denotes the real wave number, $\vec{k}^2 = k^2$.

For a *soft scatterer* one has the Dirichlet boundary condition

$$U_{|\Gamma} = 0 \tag{1.3.2}$$

whereas at infinity one requires a behavior like $e^{i\vec{k}\cdot\vec{x}}$ which corresponds to the so-called *Sommerfeld radiation conditions*

$$\begin{aligned} &U(\vec{x}) - e^{i\vec{k}\cdot\vec{x}} = O\left(|\vec{x}|^{-\frac{1}{2}}\right) \quad \text{and} \\ &\frac{\partial}{\partial|\vec{x}|}\left(U - e^{i\vec{k}\cdot\vec{x}}\right) - ik\left(U - e^{i\vec{k}\cdot\vec{x}}\right) = o\left(|\vec{x}|^{-\frac{1}{2}}\right) \quad \text{as } |\vec{x}| \to \infty. \end{aligned} \tag{1.3.3}$$

Any solution U of (1.3.1)-(1.3.3) admits the representation

$$U(\vec{x}) = e^{i\vec{k}\cdot\vec{x}} - \frac{i}{4}\int_\Gamma H_0^{(1)}\left(k|\vec{x} - \vec{\xi}|\right)\partial_n U(\vec{x})\, ds_{\vec{\xi}} \quad \text{for } \vec{x} \in \Omega^c \tag{1.3.4}$$

which depends on the still unknown second Cauchy datum

$$g = \partial_n U_{|\Gamma}.$$

Here $H_0^{(1)}(z)$ is the modified Bessel function of the first kind which is the fundamental solution of (1.3.1) satisfying radiation conditions which yield (1.3.3) for any solution of the form (1.3.4).

Taking the trace of both sides of (1.3.4) on Γ and using (1.3.2) yields the boundary integral equation of the first kind for g,

$$\frac{i}{4}\int_\Gamma g(\vec{\xi})H_0^{(1)}\left(k|\vec{x} - \vec{\xi}|\right) ds_{\vec{\xi}} = e^{i\vec{k}\cdot\vec{x}} \quad \text{for } \vec{x} \in \Gamma. \tag{1.3.5}$$

As long as k^2 is not an eigenvalue of the *interior* Dirchlet problem of the Laplacian, (1.3.5) can be solved uniquely for g and (1.3.4) will then give the desired solution.

Taking the normal derivative at Γ on both sides of (1.3.4) yields with the jump relation the Fredholm boundary integral equation of the second kind ([10] Theorem 2.19),

$$g(\vec{x}) + \frac{i}{2}\int_{\Gamma\setminus\{\vec{x}\}}\left(\partial_{n_x} H_0^{(1)}\left(k|\vec{x} - \vec{\xi}|\right)\right) g(\vec{\xi})\, ds_{\vec{\xi}} = 2i\vec{k}\cdot\vec{n}(\vec{x})e^{i\vec{k}\cdot\vec{x}} \quad \text{for } \vec{x} \in \Gamma. \tag{1.3.6}$$

This integral equation is uniquely solvable as long as k^2 is not an eigenvalue of the *interior* Neumann problem of the Laplacian.

By using additional potentials, both integral equations (1.3.5) and (1.3.6) can be modified to avoid the above-mentioned exceptional values of k^2 where one or the other is not uniquely solvable, see [10], Chapter 3.6.

If the scattering obstacle Ω is an ideally ***hard scatterer*** then one has the Neumann boundary condition

$$\partial_n U_{|\Gamma} = 0. \tag{1.3.7}$$

Here, the Green representation formula reads as

$$U(\vec{x}) = e^{i\vec{k}\cdot\vec{x}} + \frac{i}{4}\int_\Gamma U(\vec{\xi})\left(\partial_{n_{\vec{\xi}}} H_0^{(1)}(k|\vec{x}-\vec{\xi}|)\right) ds_{\vec{\xi}} \text{ for } \vec{x}\in\Omega^c; \tag{1.3.8}$$

and the unknown Cauchy datum now is

$$u := U_{|\Gamma}.$$

Taking $\vec{x}$ to Γ, we find with the jump relation for the double layer potential ([10] Theorem 2.13) the modified boundary integral equation of the second kind,

$$\begin{aligned} u(\vec{x}) &+ \frac{1}{2\pi}\int_{\vec{\xi}\in\Gamma}\left(u(\vec{\xi}) - u(\vec{x})\right) d\Theta_{\vec{x}}(\vec{\xi}) \\ &- \int_\Gamma u(\vec{\xi})\partial_{n_{\vec{\xi}}}\left\{\frac{i}{4}H_0^{(1)}\left(k|\vec{x}-\vec{\xi}|\right) + \frac{1}{2\pi}\log\left(k|\vec{x}-\vec{\xi}|\right)\right\} ds_{\vec{\xi}} = e^{i\vec{k}\cdot\vec{x}} \quad \text{for } \vec{x}\in\Gamma. \end{aligned} \tag{1.3.9}$$

Again, (1.3.9) is uniquely solvable as long as k^2 is not an eigenvalue of the interior Dirchlet problem of the Laplacian. The behavior of (1.3.9) for domains including corners was studied in [49].

If we take the normal derivative on both sides of (1.3.8) then we obtain the hypersingular boundary integral equation of the first kind,

$$\frac{i}{4}\partial_{n_{\vec{x}}}\int_\Gamma u(\vec{\xi})\partial_{n_{\vec{\xi}}} H_0^{(1)}\left(k|\vec{x}-\vec{\xi}|\right) ds_{\vec{\xi}} = -\partial_{n_{\vec{x}}} e^{i\vec{k}\cdot\vec{x}} \quad \text{for } \vec{x}\in\Gamma, \tag{1.3.10}$$

together with the compatibility condition

$$\int_\Gamma u\, ds = 0.$$

This hypersingular equation was analyzed by Filippi and by Giroire and Nedelec (see [24] and [29]). It is uniquely solvable if k^2 is not an eigenvalue of the interior Neumann problem for the Laplacian.

1.4 Some problems of elastostatics

Elastostatic problems have been treated rather early by means of boundary integral equations and boundary element methods, see e.g. the books [54], [64], [55], [7], [48].

The displacement field $\vec{u}(\vec{x})$ of an ideal elastic homogeneous and isotropic material is governed by the ***Navier equations***

$$\mu \, \Delta \, \vec{u} + (\lambda + \mu) \, \mathrm{grad} \, \mathrm{div} \, \vec{u} = \vec{0} \tag{1.4.1}$$

in Ω for interior poblems, or in Ω^c for exterior problems, respectively. λ and μ denote the Lamé constants with $\lambda > -\mu$ and $\mu > 0$ for two-dimensional and $\lambda > -\frac{2}{3}\mu$ and $\mu > 0$ for three-dimensional problems. Any solution of (1.4.1) in Ω can be represented by the Betti formula in terms of elastic single- and double-layer potentials,

$$\vec{u}(\vec{x}) = \int_\Gamma \left\{ \underset{\sim}{F}(\vec{x},\vec{\xi})\vec{t}(\vec{\xi}) - \left(T_{\vec{\xi}}\underset{\sim}{F}(\vec{x},\vec{\xi})\right)^T \vec{u}(\vec{\xi}) \right\} ds_{\vec{\xi}} \quad \text{for } \vec{x} \in \Omega. \tag{1.4.2}$$

Here $\underset{\sim}{F}$ denotes the Kelvin fundamental matrix for (1.4.1), given by

$$\underset{\sim}{F}(\vec{x},\vec{\xi}) = \frac{\lambda + 3\mu}{4\pi(\lambda + 2\mu)} \left\{ -\log|\vec{x} - \vec{\xi}|\mathbf{1} + \frac{(\vec{x} - \vec{\xi})(\vec{x} - \vec{\xi})^T}{|\vec{x} - \vec{\xi}|^2} \right\}. \tag{1.4.3}$$

$\mathbf{1}$ denotes the unity matrix. $T_{\vec{\xi}}$ denotes the traction operator on Γ with differentiations with respect to $\vec{\xi}$, in two dimensions given by

$$\vec{t} = T\vec{u}_{|\Gamma} := \lambda(\mathrm{div}\,\vec{u}) + 2\mu\partial_n\vec{u} - \mu\left(\frac{\partial u_2}{\partial x_1} - \frac{\partial u_1}{\partial x_2}\right)\frac{d\vec{\xi}}{ds} \tag{1.4.4}$$

where $\vec{t}$ denotes the boundary traction. In $T_{\vec{\xi}}\underset{\sim}{F}(\vec{x},\vec{\xi})$, the traction operator operates on each of the columns of $\underset{\sim}{F}$ separately.

For (1.4.1), in the *Dirichlet problem*, the displacement is given on the boundary,

$$\vec{u}_{|\Gamma} = \vec{\varphi}. \tag{1.4.5}$$

Now, the Cauchy data are $(\vec{u},\vec{t})^T$ and for the displacement problem, $\vec{t}$ is the missing Cauchy datum.

If $\vec{x}$ in (1.4.2) is taken to the boundary Γ, then one obtains the Fredholm boundary integral equation of the first kind for $\vec{t}$,

$$\int_\Gamma \underset{\sim}{F}(\vec{x},\vec{\xi})\vec{t}(\vec{\xi})\,ds_{\vec{\xi}} = \frac{1}{2}\vec{\varphi}(\vec{x}) + \int_{\Gamma\setminus\{\vec{x}\}} \left(T_{\vec{\xi}}\underset{\sim}{F}(\vec{x},\vec{\xi})\right)^T \vec{\varphi}(\vec{\xi})\,ds_{\vec{\xi}} \quad \text{for } \vec{x} \in \Gamma, \tag{1.4.6}$$

provided Γ is smooth. If Γ has corner points, then the given right-hand side in (1.4.6) needs to be modified accordingly [34].

If we apply the traction operator $T_{\vec{x}}$ to both sides of (1.4.3) then we find the boundary integral equation

$$\begin{aligned} &\vec{t}(\vec{x}) - 2\int_{\Gamma\setminus\{\vec{x}\}} \left(T_{\vec{x}}\underset{\sim}{F}(\vec{x},\vec{\xi})\right)\vec{t}(\vec{\xi})\,ds_{\vec{\xi}} \\ &\qquad = -2T_{\vec{x}}\int_{\Gamma\setminus\{\vec{x}\}} \left(T_{\vec{\xi}}\underset{\sim}{F}(\vec{x},\vec{\xi})\right)^T \vec{\varphi}(\vec{\xi})\,ds_{\vec{\xi}} \quad \text{for } \vec{x} \in \Gamma. \end{aligned} \tag{1.4.7}$$

A closer analysis of the kernel $\left(T_{\vec{\xi}}\underset{\sim}{F}(\vec{x},\vec{\xi})\right)^T$ and its adjoint $T_{\vec{x}}\underset{\sim}{F}(\vec{x},\vec{\xi})$ shows that both are Cauchy singular. Hence, the left-hand side of (1.4.7) is given by a *Cauchy singular integral operator*. As is shown in [54], (1.4.7) admits exactly one solution $\vec{t}$ for any given $\vec{\varphi}$.

The Neumann problem corresponds here to the *traction boundary condition* where

$$\vec{t} = T\vec{u}_{|\Gamma} = \vec{\psi} \quad \text{on } \Gamma \tag{1.4.8}$$

is given. For the interior problem, $\vec{t}$ in (1.4.8) must satisfy the equilibrium conditions

$$\int_\Gamma \vec{t}\, ds = \vec{0} \text{ and } \int_\Gamma (t_1\xi_2 - t_2\xi_1)\, ds_{\vec{\xi}} = 0. \tag{1.4.9}$$

For $\vec{x}$ taken to the boundary Γ, the representation formula (1.4.3) yields the *Cauchy singular integral equation* for the missing Cauchy datum $\vec{u}_{|\Gamma}$,

$$\begin{aligned}\vec{u}(\vec{x}) + 2\int_{\Gamma\setminus\{\vec{x}\}} &\left(T_{\vec{\xi}}\underset{\sim}{F}(\vec{x},\vec{\xi})\right)^T \vec{u}(\vec{\xi})\, ds_{\vec{\xi}} \\ &= 2\int_\Gamma F(\vec{x},\vec{\xi})\vec{\psi}(\vec{\xi})\, ds_{\vec{\xi}} \quad \text{for } \vec{x}\in\Gamma.\end{aligned} \tag{1.4.10}$$

Since for rigid motions $\vec{u}_0 = \vec{a} + b(-x_2, x_1)^T$ with constant b and $\vec{a}$, the traction vanishes, $T\vec{u}_0 = \vec{0}$, the boundary integral equation (1.4.10) has $\vec{u}_0$ as an eigenspace, and it can be shown that these are all eigensolutions. Moreover, for given $\vec{\psi}$ with (1.4.9), the right-hand side in (1.4.10) satisfies the required orthogonality conditions and the solution $\vec{u}$ to (1.4.10) becomes unique if we append the additional conditions

$$\int_\Gamma \vec{u}\, ds = \vec{0} \text{ and } \int_\Gamma (u_1\xi_2 - u_2\xi_1)\, ds_{\vec{\xi}} = 0. \tag{1.4.11}$$

If we take the boundary traction of both sides in (1.4.3) then we find for $\vec{u}_{|\Gamma}$ the *hypersingular integral equation* of the first kind,

$$\begin{aligned}-T_{\vec{x}}\int_\Gamma &\left(T_{\vec{\xi}}\underset{\sim}{F}(\vec{x},\vec{\xi})\right)^T \vec{u}(\vec{\xi})\, ds_{\vec{\xi}} \\ &= \frac{1}{2}\vec{\psi}(\vec{x}) - \int_{\Gamma\setminus\{\vec{x}\}} \left(T_{\vec{x}}\underset{\sim}{F}(\vec{x},\vec{\xi})\right) \vec{\psi}(\vec{\xi})\, ds_{\vec{\xi}} \quad \text{for } \vec{x}\in\Gamma.\end{aligned} \tag{1.4.12}$$

Also here, the rigid motions $\vec{u}_0$ are the eigensolutions of (1.4.12) and the solution becomes unique if we require (1.4.11), too.

Hence, in case of elastostatics, we again find three different types of boundary integral equations:

- the Fredholm integral equation of the first kind (1.4.6) with weakly singular kernel,
- the Cauchy singular integral equations (1.4.7) and (1.4.10),
- the hypersingular equation of the first kind (1.4.12).

1.5 The boundary integral equations of the direct approach for general elliptic boundary value problems of even order

Following [18] let us now generalize the previous analysis to general elliptic boundary value problems. Let us consider the linear partial differential equations of order $2m$,

$$P^{(2m)}U = f \quad \text{in } \Omega \text{ (or } \Omega^c\text{, respectively).} \tag{1.5.1}$$

Let

$$R\gamma U = \varphi \quad \text{on } \Gamma \tag{1.5.2}$$

denote the boundary conditions. The unknown U stands in general for a vector-valued function. $R = ((R_{jk}))$ is a given matrix of tangential operators on Γ, $j = 0, \dots, m-1$; $k = 0, \dots, 2m-1$, of orders $\mu_j - k$, $0 \le \mu_j \le 2m-1$, $R_{jk} = 0$ for $\mu_j < k$.

$$\gamma U := \left(u, \partial_n U, \dots, \partial_n^{2m-1} U\right)^T_{|\Gamma} \tag{1.5.3}$$

denotes the trace operator on Γ, hence γU is the set of Cauchy-data, too. For the representation formula we need the decomposition

$$P^{(2m)} = \sum_{j=0}^{2m} P_j \partial_n^j \quad \text{along } \Gamma \tag{1.5.4}$$

with tangential operators P_j of orders $2m - j$. Let us assume the existence of a fundamental solution $\underset{\sim}{F}(\vec{x}, \vec{y})$ satisfying the adjoint differential equation

$$P_{\vec{y}}^{(2m)\prime} \tilde{F}(\vec{x}, \vec{y}) = \delta_{\vec{x}}(\vec{y}), \tag{1.5.5}$$

then we have the representation formula

$$U(\vec{x}) = \int_\Omega \underset{\sim}{F}(\vec{x}, \vec{\xi})\, d\xi \mp \sum_{k=0}^{2m-1} \sum_{l=0}^{2m-1-k} \int_\Gamma \left(\partial_{n\vec{\xi}}^{\prime k} \underset{\sim}{F}(\vec{x}, \vec{\xi})\right) P_{k+l+1} \partial_n^l U(\vec{\xi})\, ds_{\vec{\xi}} \tag{1.5.6}$$

for $\vec{x} \in \Omega$ with the $-$ sign; and for $\vec{x} \in \Omega^c$ with the $+$ sign. ∂_n' is the differential operator adjoint to ∂_n.

We further assume that the operators in R admit the existence of a complementary boundary condition S such that the matrix of boundary operators $M := \begin{pmatrix} R \\ S \end{pmatrix}$ is invertible. Thus, we define $w = S\gamma U$, and

$$\begin{pmatrix} \varphi \\ w \end{pmatrix} = M\gamma U = \begin{pmatrix} R \\ S \end{pmatrix} U, \quad \gamma U = M^{-1} \begin{pmatrix} \varphi \\ w \end{pmatrix}, \tag{1.5.7}$$

can be replaced in (1.5.6). Then application of M to both sides of (1.5.6) yields the overdetermined system of equations

$$(1.5.8)\qquad \begin{pmatrix} \varphi \\ w \end{pmatrix} = \begin{pmatrix} R \\ S \end{pmatrix} \gamma \Big\{ \int_\Omega \underset{\sim}{F} f \, d\xi \mp \sum_{k=0}^{2m-1} \sum_{l=0}^{2m-1-k} \int_\Gamma \big(\partial'^{k}_{n\vec\xi} \underset{\sim}{F}(\vec x, \vec\xi) \big) P_{k+l+1} M_l^{-1} \begin{pmatrix} \varphi \\ w \end{pmatrix} ds_{\vec\xi} \Big\}$$

for the missing Cauchy datum w. M_l^{-1} denotes the l-th row of M^{-1}. Clearly, if M is the identity matrix then (1.5.8) is just the Calderon projector; and in general, (1.5.8) is its extension to invertible M.

Now,

$$(1.5.9)\qquad Aw := \mp \sum_{k=0}^{2m-1} \sum_{l=0}^{2m-1-k} R\gamma \int_\Gamma \big(\partial'^{k}_{n\vec\xi} \underset{\sim}{F}(\vec x, \vec\xi) \big) P_{k+l+1} M_l^{-1} \begin{pmatrix} 0 \\ w \end{pmatrix} ds_{\vec\xi}$$

is a matrix of operators corresponding to the boundary integral equations of the first kind, whereas

$$(1.5.10)\quad Bw := w \pm \sum_{k=0}^{2m-1} \sum_{l=0}^{2m-1-k} S\gamma \int_\Gamma \big(\partial'^{k}_{n\vec\xi} \underset{\sim}{F}(\vec x, \vec\xi) \big) P_{k+l+1} M_l^{-1} \begin{pmatrix} 0 \\ w \end{pmatrix} ds_{\vec\xi} \text{ on } \Gamma$$

corresponds to boundary integral equations of the second kind; namely for given φ, the first equation of (1.5.8) with (1.5.9) is a boundary integral equation of the first kind for w; the second integral equation of (1.5.8) is a boundary integral equation of the second kind for w on Γ.

Clearly, if we specialize $P^{(2m)}$, R, and S, all the boundary integral equations in the previous sections are special cases of (1.5.9) or of (1.5.10), respectively.

2 The Characterization of Boundary Integral Operators and Galerkin Boundary Element Methods

As we have seen in the previous Sections, all the boundary integral operators generated by boundary value problems are compositions of tangential differential operators and boundary potentials defined by normal derivatives of the fundamental solution, its boundary traces including normal derivatives on Γ and, again, tangential differential operators. This yields the following basic result on joint properties of all boundary integral operators on Γ, provided $\Gamma \in C^{\infty}$.

Theorem 2.1 *The boundary integral operators associated with regular elliptic boundary value problems are classical pseudo-differential operators of integer orders. That means in particular, that (after appropriate integrations by parts) they can be represented as compositions of tangential derivatives and Cauchy singular integral operators or, correspondingly, with inverse derivatives of the Cauchy kernel.*

The proof of this Theorem follows from [21] 23.43.14.3 and [23] Section I§3, and will be presented in detail in [88]. For the definition of pseudo-differential operators see e.g. [95]. In the following we shall draw some conclusions from this common property.

2.1 The representation and the order of boundary integral operators

More generally than in the previous chapters, let us consider here boundary value problems in n dimensions with $n = 2$ or 3, respectively. Let $\vec{x} \in \Gamma$ be chosen arbitrarily and fixed, for the time being. The boundary curve or surface, respectively, can be represented by

$$\vec{\xi} = \vec{x} + \Phi(\vec{x};\vec{t})\vec{n}(\vec{x}) + \sum_{j=1}^{n-1} t_j \vec{e}_j(\vec{x}) \in \Gamma \tag{2.1.1}$$

where $\vec{t} = (t_1, t_{n-1})^T$ with $n = 2$ or 3 are the parameters, where $\vec{e}_j(\vec{x})$, $j = 1, n-1$, denote some chosen basis of tangential vectors to Γ at $\vec{x}$, and where Φ is a smooth function describing the distance from the tangential plane to Γ satisfying

$$\Phi(\vec{x};0) = 0, \ \frac{\partial \Phi}{\partial t_j}(\vec{x},0) = 0, \ \ j = 1, n-1.$$

Using (2.1.1), we define

$$w(\vec{x};\vec{t}) := w\left(\vec{\xi}(\vec{x},\vec{t})\right) \tag{2.1.2}$$

in the neigborhood of $\vec{x}$. As a consequence of Theorem 2.1, we have:

Theorem 2.2 *Let A be any of the boundary integral operators on $\Gamma \in C^\infty$ associated with a regular elliptic boundary value problem. Then A admits the representation*

$$\begin{aligned} Aw(\vec{x}) &= \sum_{|l|\le\alpha} \Big\{ b_l(\vec{x}) \partial_t^l w(\vec{x};\vec{t})\big|_{\vec{t}=0} \\ &+ \sum_{p\ge 0} \text{p.v.} \int_{|\vec{t}|>0} \Big\{ |\vec{t}|^{p-n+1} f_{1lp}\left(\vec{x};\frac{\vec{t}}{|\vec{t}|}\right) + |\vec{t}|^p \log|\vec{t}| f_{2lp}\left(\vec{x};\frac{\vec{t}}{|\vec{t}|}\right)\Big\} \chi(|\vec{t}|)\partial_t^l w\, dt \Big\} \\ &+ \int_\Gamma k(\vec{\xi},\vec{x}) w(\vec{\xi})\, ds_{\vec{\xi}} \end{aligned} \tag{2.1.3}$$

for $\alpha \ge 0$ or

$$\begin{aligned} Aw(\vec{x}) &= \sum_{p\ge 0} \int_{|\vec{t}|>0} |\vec{t}|^{p-\alpha-n+1} f_{1p}\left(\vec{x};\frac{\vec{t}}{|\vec{t}|}\right) \chi(|\vec{t}|) w(\vec{x};\vec{t})\, dt \\ &+ \sum_{p\ge 0 \wedge p \ge -\alpha-n+1} \int_{|\vec{t}|>0} |\vec{t}|^p \log|\vec{t}| f_{2p}\left(\vec{x};\frac{\vec{t}}{|\vec{t}|}\right) \chi(|\vec{t}|) w(\vec{x};\vec{t})\, dt \\ &+ \int_\Gamma k(\vec{\xi},\vec{x}) w(\vec{\xi})\, ds_{\vec{\xi}} \end{aligned} \tag{2.1.4}$$

for $\alpha < 0$.

α is the order of A, which here is either an integer, or we set $\alpha = -\infty$ if all terms except k vanish indentically and $k \in C^\infty(\Gamma \times \Gamma)$.

For $\alpha \in \mathbb{N}_0$, the characteristic functions f_{1l0} satisfy the *Mikhlin conditions*

$$\begin{aligned} &f_{1l0}(\vec{x};1) + f_{1l0}(\vec{x};-1) = 0 \ \text{ for } n = 2 \text{ and } |l| = l \le \alpha; \\ &\int_{\varphi=0}^{2\pi} f_{1l0}\left(\vec{x};(\cos\varphi,\sin\varphi)^T\right) d\varphi = 0 \ \text{ for } n = 3 \\ &\text{and } |l| = |(l_1,l_2)| = l_1 + l_2 \le \alpha, \ \ l \in \mathbb{N}_0^2. \end{aligned} \tag{2.1.5}$$

For given A and given local representation (2.1.1) of Γ, the order α, the coefficients $b_l(\vec{x})$, and the characteristic functions f_{1lp}, f_{2lp} are uniquely determined. $\partial_t^l = \partial_{t_1}^{l_1}\partial_{t_2}^{l_2}$. Here $\chi(\rho)$ is a C^∞ cut-off function of one variable with $\chi \equiv 1$ in a neighborhood of 0 and with a compact support which is small enough so that (2.1.1) is valid for all $|\vec{t}| \in \operatorname{supp}\chi$. The sums in (2.1.3) and (2.1.4), respectively, are understood asymptotically in the following sense: They are valid for $p \le N \in \mathbb{N}$ for any chosen N. Then the kernel k of the remainder is in $C^{N-n+1}(\Gamma \times \Gamma)$.

The proof of Theorem 2.2 is based on the representation of pseudo-differential operators with pseudo-homogeneous kernels, see [89] Chapter III; a detailed proof will be given in [88].

In the tangential coordinates, the representations (2.1.3) and (2.1.4) can be Fourier transformed. The Fourier inversion formula yields the more familiar form of pseudo-differential operators as follows.

Lemma 2.3 *The operator A given as above can be written as*

$$
\begin{aligned}
Aw(\vec{x}) &= \int_{\mathbf{R}^{n-1}} \int_{\mathbf{R}^{n-1}} e^{-2\pi i \vec{\eta}\cdot\vec{\tau}} \left(a(\vec{x};\vec{\eta})\chi(|\vec{\tau}|)\right) w(\vec{x};\vec{\tau})\, d\tau\, d\eta \\
&\quad + \int_\Gamma k^*(\vec{\xi},\vec{x}) w(\vec{\xi})\, ds_{\vec{\xi}}.
\end{aligned} \tag{2.1.6}
$$

Here the so-called symbol $a(\vec{x};\vec{\eta})$ is given by

$$
a(\vec{x};\vec{\eta}) = A\left(e^{2\pi i\vec{\eta}\cdot\bullet}\chi(|\bullet|)\right)(\vec{x}). \tag{2.1.7}
$$

According to the asymptotic representations (2.1.3) or (2.1.4), respectively, also the symbol a can be written in the asymptotic form

$$
\begin{aligned}
a(\vec{x};\vec{\eta}) &= \sum_{|l|\le\alpha} \Big\{ b_l(\vec{x})(i\vec{\eta})^l + (i\vec{\eta})^l \sum_{p\ge 0} p.f. \int_{|\vec{t}|>0} e^{2\pi i\vec{t}\cdot\vec{\eta}} \left\{ |\vec{t}|^{p-n+1} f_{1lp} \right. \\
&\quad \left. + |\vec{t}|^p \log|\vec{t}| f_{2lp} \right\} \chi(|\vec{t}|\, dt \Big\} + \int_{\mathbf{R}^{n-1}} k\left(\vec{\xi}(\vec{t});\vec{x}\right) e^{2\pi i\vec{\eta}\cdot\vec{t}} \chi(|\vec{t}|)\, dt
\end{aligned} \tag{2.1.8}
$$

for $\alpha \ge 0$ and

$$
\begin{aligned}
a(\vec{x};\vec{\eta}) &= \sum_{p\ge 0} \int_{|\vec{t}|>0} e^{2\pi i\vec{t}\cdot\vec{\eta}} |\vec{t}|^{p-\alpha-n+1} f_{1p}\chi(|\vec{t}|)\, dt \\
&\quad + \sum_{p\ge 0 \wedge p \ge -\alpha-n+1} \int_{|\vec{t}|>0} e^{2\pi i\vec{t}\cdot\vec{\eta}} |\vec{t}|^p \log|\vec{t}| f_{2p}\chi(|\vec{t}|)\, dt \\
&\quad \int_{\mathbf{R}^{n-1}} k\left(\vec{\xi}(\vec{t});\vec{x}\right) e^{2\pi i\vec{\eta}\cdot\vec{t}} \chi(|\vec{t}|)\, dt
\end{aligned} \tag{2.1.9}
$$

for $\alpha < 0$.

Again, (2.1.8) and (2.1.9) are to be understood asymptotically.

Because of Theorem 2.1, the symbol a of A is a 'classical' symbol which can be characterized in the following way. We consider the classical classes of symbols

$$\begin{aligned} \mathcal{S}^j_{1,0} &:= \Big\{ q \in C^\infty(\Gamma \times \mathbb{R}^{n-1}) \Big| \forall\, k, l \in \mathbb{N}_0^{n-1} \exists c_{lk} \geq 0 : \\ &\qquad \Big| \partial^k_{\vec{x}} \partial^l_{\vec{\eta}} q(\vec{x};\vec{\eta}) \Big| \leq c_{lk} (1 + |\vec{\eta}|)^{j-|l|} \Big\}. \end{aligned} \tag{2.1.10}$$

Then a has the asymptotic expansion

$$a(\vec{x};\vec{\eta}) \simeq \sum_{p \geq 0} a_p(\vec{x};\vec{\eta}) \qquad \text{with } a_p \in \mathcal{S}^{\alpha-p}_{1,0}, \text{ i.e.} \tag{2.1.11}$$

$$a(\vec{x};\vec{\eta}) - \sum_{p=0}^{N} a_p(\vec{x};\vec{\eta}) \in \mathcal{S}^{\alpha-(N+1)}_{1,0} \qquad \text{for every } N \in \mathbb{N}_0, \tag{2.1.12}$$

$$a_p(\vec{x};\lambda\vec{\eta}) = \lambda^{\alpha-p} a_p(\vec{x};\vec{\eta}) \qquad \text{for } \lambda \geq 1,\ |\vec{\eta}| \geq 1. \tag{2.1.13}$$

Here α is the order of A, again. In fact, the terms on the right-hand side of (2.1.7) or (2.1.9), respectively, correspond to the terms a_p of the symbol in (2.1.11) modulo C^∞-functions with compact support with respect to $\vec{\eta}$.

Definition: $a_0(\vec{x};\vec{\eta})$ is called the *principal symbol* of A.

It should be noted that the terms in the asymptotic expansion (2.1.11) are uniquely defined for $|\vec{\eta}| \geq 1$. The classical symbol classes have been introduced by Kohn and Nirenberg [50], see also [95].

For $n = 2$ and a C^∞ simple Jordan curve Γ, we can use a global parametric representation with some 1-periodic 2-vector-valued function $\vec{\Xi}(t)$ and

$$\vec{\xi} = \vec{\Xi}(t), \quad \vec{x} = \vec{\Xi}(\tau) = \vec{\Xi}(\tau \pm 1), \tag{2.1.14}$$

instead of the local representation (2.1.1). Then all functions on Γ correspond to 1-periodic functions which can be replaced by their Fourier series. In this case, due to Agranovič [2], the representation of the boundary integral operators in question can be rewritten into a rather handy form,

$$A\left(\sum_{l\in\mathbb{Z}} \hat{w}_l e^{2\pi i l \bullet}\right)(\tau) = \sum_{l\in\mathbb{Z}} a(\vec{x};l)\hat{w}(l)e^{2\pi i l\tau} + \int_0^1 k(\vec{x};t)w(t)\,dt \tag{2.1.15}$$

where a is the symbol of A and $k \in C^\infty$. With (2.1.11)-(2.1.13) we can also write

$$A\left(\sum_{l\in\mathbb{Z}} \hat{w}_l e^{2\pi i l \bullet}\right)(\tau) = \sum_{l\in\mathbb{Z}\setminus\{0\}} |l|^\alpha a_0(\vec{x};\operatorname{sign} l)\hat{w}(l)e^{2\pi i l\tau} + A_{\alpha-1}w(\tau) \tag{2.1.16}$$

where

$$a_0(\vec{x};l) = |l|^\alpha a_0(\vec{x};\operatorname{sign} l) \quad \text{for } l \neq 0 \tag{2.1.17}$$

is the principal symbol of A, and where the remaining integral operator $A_{\alpha-1}$ is a pseudo-differential operator of order $\alpha - 1$. For proofs of (2.1.15) and (2.1.16) see [59] and [80].

For illustration let us classify the operators from Chapter 1 and list the corresponding principal symbols a_0 for $|\eta| \geq 1$.

$\underline{n=2}$: **Operators of order** 0:

$$a_0 = 1 : (1.1.18), (1.1.31), (1.3.6), (1.3.9) \tag{2.1.18}$$

$$a_0(\vec{x};\eta) = \begin{pmatrix} 1 & , & -i\gamma\frac{\eta}{|\eta|} \\ i\gamma\frac{\eta}{|\eta|} & , & 1 \end{pmatrix} \quad \text{where } 0 < \gamma = \tfrac{\mu}{\lambda+2\mu} < 1 : \ (1.4.7), (1.4.10); \tag{2.1.19}$$

Operators of order -1:

$$a_0 = \frac{1}{|\eta|} : \ (1.1.29), (1.2.9), (1.3.5), (1.4.6); \tag{2.1.20}$$

Operators of order $+1$:

$$a_0 = c_0|\eta|, \ c_0 \neq 0 : \ (1.1.22), (1.3.10), (1.4.12). \tag{2.1.21}$$

For $\underline{n=3}$, we find the same principal symbols for the operators corresponding to (1.1.18), (1.1.31), (1.3.6), (1.3.9), (1.1.29), (1.3.5), (1.1.22), (1.3.10). In elasticity and for Stokes flows, the equations corresponding to (1.4.7), (1.4.10) and to the weakly singular and hypersingular cases are more complicated (see e.g. [97] and [25]). Here let us list the corresponding principal symbols. Let $\underset{\sim}{M}(x) := (\vec{e}_1(\vec{x}), \vec{e}_2(\vec{x}), \vec{n}(\vec{x}))^T$.

Operators of order 0:

$$a_0(\vec{x};\vec{\eta}) = \underset{\sim}{M}(\vec{x})^{-1}\frac{1}{|\vec{\eta}|}\begin{pmatrix} |\vec{\eta}| & , & 0 & , & -i\gamma\eta_1 \\ 0 & , & |\vec{\eta}| & , & -i\gamma\eta_2 \\ i\gamma\eta_1 & , & i\gamma\eta_2 & , & |\vec{\eta}| \end{pmatrix}\underset{\sim}{M}(\vec{x}) \tag{2.1.22}$$

equations (1.4.7) and (1.4.10) (see [62] IV §6)).

Operators of order -1:

$$a_0(\vec{x};\vec{\eta}) = c_0\underset{\sim}{M}(\vec{x})^{-1}\frac{1}{|\vec{\eta}|^3}\begin{pmatrix} |\vec{\eta}|^2 + \kappa\eta_2^2 & , & -\kappa\eta_1\eta_2 & , & 0 \\ -\kappa\eta_1\eta_2 & , & |\vec{\eta}|^2 + \kappa\eta_1^2 & , & 0 \\ 0 & , & 0 & , & |\vec{\eta}|^2 \end{pmatrix}\underset{\sim}{M}(\vec{x}) \tag{2.1.23}$$

in elasticity equations corresponding to (1.4.6) with $c_0 = \frac{\lambda+3\mu}{2\mu(\lambda+2\mu)}$ and $0 < \kappa = \frac{\lambda+\mu}{\lambda+3\mu} < 1$ [45], [97] and in viscous flows equations corresponding to (1.2.9) with $c_0 = -\frac{\pi}{4}$ and $\kappa = 1$ [25].

Operators of order +1:

$$(2.1.24)\quad a_0(\vec{x};\vec{\eta}) = c_0 \underset{\sim}{M}(\vec{x})^{-1}\frac{1}{|\vec{\eta}|}\begin{pmatrix} |\vec{\eta}|^2+\varepsilon\eta_1^2 \, , & \varepsilon\eta_1\eta_2 \, , & 0 \\ \varepsilon\eta_1\eta_2 \, , & |\vec{\eta}|^2+\varepsilon\eta_2^2 \, , & 0 \\ 0 \, , & 0 \, , & (1+\varepsilon)|\vec{\eta}|^2 \end{pmatrix}\underset{\sim}{M}(\vec{x})$$

in elasticity equations corresponding to (1.4.12) with $c_0 = \mu^2$ and $\varepsilon = \frac{\lambda}{\lambda+2\mu}$ [45], [97].

Of course, also the operators A in (1.5.9) and B in (1.5.10) are defined by matrices $((A_{jk}))\, ,((B_{jk}))$ of classical pseudo-differential operators on Γ. In [18] the corresponding orders are given by

$$\text{ord } A_{jk} = \mu_j + \mu_k + 1 - 2m \quad \text{and}$$
$$\text{ord } B_{jk} = \mu_j - \mu_k .$$

2.2 Variational formulation and strong ellipticity

As in finite element methods for elliptic boundary value problems, also in boundary element methods, the variational formulation and coerciveness properties provide the basic mathematical foundation for rigorous error and convergence analysis. The variational formulation rests on the weak formulation of boundary integral equations on Γ, which in engineering analysis is also called the ***weighted residual technique***. More explicitly, if the boundary integral equation

$$(2.2.1)\qquad Au = f \quad \text{on } \Gamma$$

is to be solved, then the corresponding weak formulation reads as to find u such that for all test functions w there holds on Γ

$$(2.2.2)\qquad \int_\Gamma (Au)\bar{w}\, ds = \int_\Gamma f\bar{w}\, ds.$$

For further analysis of the bilinear form on the left-hand side of (2.2.2) and the mapping properties involved, we introduce Sobolev spaces of functions on Γ associated with the scalar products

$$(2.2.3)\qquad (f,g)_{H^\sigma(\Gamma)} := \sum_{k\in\mathbb{Z}} |\delta_{0k}+k|^{2\sigma}\, \hat{f}_k \quad \text{for } n=2$$

and $\sigma \in \mathbb{R}$, and

$$(2.2.4)\qquad \begin{aligned} \langle f,g\rangle_{H^\sigma(\Gamma)} := & \sum_{|l|\le[\sigma]}\sum_{j=1}^{M}\int_{\Gamma_j} \partial_t^l f \overline{\partial_t^l g}\, ds \\ & +(1-\delta_{\sigma,[\sigma]})\sum_{|l|=[\sigma]}\sum_{j=1}^{M}\int_{\Gamma_j}\int_{\Gamma_j} \frac{\left(\partial_t^l f(\vec{x}) - \partial_t^l f(\vec{y})\right)\left(\partial_t^l g(\vec{x}) - \partial_t^l g(\vec{x})\right)}{|\vec{x}-\vec{y}|^{n-1+2(\sigma-[\sigma])}}\, ds_{\vec{x}} ds_{\vec{y}} \end{aligned}$$

for $n \geq 2$ and $\sigma \geq 0$. Here ∂_t^l denote all covariant derivatives of order $|l|$ in Γ and $\bigcup_{j=1}^{M} \Gamma_j$ is a finite covering of Γ by regular charts. $[\bullet]$ denotes the Gaussian bracket. Then the norms

$$\|f\|_{H^\sigma(\Gamma)} = (f, f)^{\frac{1}{2}}_{H^\sigma(\Gamma)} \tag{2.2.5}$$

and

$$\|f\|_{H^\sigma(\Gamma)} = \langle f, f\rangle^{\frac{1}{2}}_{H^\sigma(\Gamma)} \tag{2.2.6}$$

turn out to be equivalent [59]; and the norms belonging to different finite coverings of Γ by regular charts are equivalent, too. The Sobolev space $H^\sigma(\Gamma)$ on the regular C^∞ manifold Γ can be defined by the closure of $C^\infty(\Gamma)$ with respect to (one of) the norms $\|\cdot\|_{H^\sigma(\Gamma)}$. Although in general the norms in (2.2.5) and (2.2.6) are different, we shall in the following still use the same notation, since here one of them can always be replaced by the other one.

For $\sigma < 0$ we use the duality with respect to $L^2(\Gamma)$,

$$\|f\|_{H^\sigma(\Gamma)} = \sup_{\|\phi\|_{H^{-\sigma}(\Gamma)} \leq 1} \left|\langle f, \phi\rangle_{L^2(\Gamma)}\right| \tag{2.2.7}$$

where

$$\langle f, \phi\rangle_{L^2(\Gamma)} := \int_\Gamma f\bar{\phi}\, ds. \tag{2.2.8}$$

Since the boundary integral operators considered here are also pseudo-differential operators we have the following mapping property.

Theorem 2.4 ([37]) *For a pseudo-differential operator A of order $\alpha \in \mathbb{R}$ and any $\sigma \in \mathbb{R}$, the mapping*

$$A : H^\sigma(\Gamma) \to H^{\sigma-\alpha}(\Gamma) \tag{2.2.9}$$

is continuous.

For the proof see [37] or [95] II §6.

Since we are interested in solving (2.2.1), the continuity (2.2.9) is not enough; we also need properties that provide the existence of A^{-1}. The corresponding basic property is *strong ellipticity*, defined as follows.

Definition: A $p \times p$ matrix of pseudo-differential operators A of order α is called *strongly elliptic* iff there exist a positive constant γ_0 and a complex matrix-valued function $\Theta \in C^\infty(\Gamma)$ such that for all $\vec{x} \in \Gamma$ and all $\vec{\eta} \in \mathbb{R}^{n-1}$ with $|\vec{\eta}| = 1$ and all $\zeta \in \mathbb{C}^p$ the principal symbol of A satisfies

$$\operatorname{Re} \zeta^T \left(\Theta(\vec{x}) a_0(\vec{x}; \vec{\eta})\right) \bar{\zeta} \geq \gamma_0 |\zeta|^2. \tag{2.2.10}$$

Lemma 2.5 ([70]) *In case $n = 2$, strong ellipticity* (2.2.10) *is equivalent to the property:*
For all $\vec{x} \in \Gamma$ and all $\mu \in [0,1]$, the principal symbol a_0 satisfies

$$\text{Det}\,\{\mu a_0(\vec{x};1) + (1-\mu)a_0(\vec{x};-1)\} \neq 0. \tag{2.2.11}$$

For proofs see [70], [71], and [72].

The proposed coerciveness result reads as follows.

Theorem 2.6 ([37], [50]) *If A is a strongly elliptic $p \times p$ system of pseudo-differential operators of order α then there exists a positive constant γ_1 and a completely continuous operator $C : H^{\frac{\alpha}{2}}(\Gamma) \to H^{\frac{\alpha}{2}}(\Gamma)$ such that for all $w \in H^{\frac{\alpha}{2}}(\Gamma)$*

$$\text{Re}\,\langle(\Theta A + C)w, w\rangle_{L^2(\Gamma)} \geq \gamma_1 \|w\|^2_{H^{\frac{\alpha}{2}}(\Gamma)}. \tag{2.2.12}$$

In other words, $D = \Theta A + C$ is $H^{\frac{\alpha}{2}}(\Gamma)$-elliptic on Γ.

The proof of (2.2.12) based on (2.2.10) can also be found in [95] II.§8.

Remarks: Note, that here strong ellipticity of A in (2.2.1) is invariant under multiplications of the equation (2.2.1) with invertible Θ, which is without significance for the equation. For the various boundary element methods applied to (2.2.1), however, the invariant definition is significant. Moreover, Lemma 2.5 is valid only with the Definition (2.2.10). The concept of strong ellipticity and Gårding's inequality (2.2.12) can be generalized to systems with operators of different orders [98].

Note that all the special operators listed in Chapter 1 with the principal symbols (2.1.19)-(2.1.24) are strongly elliptic with $\Theta = c_0$ or $\Theta = 1$.

Many elliptic boundary value problems can be reformulated in terms of variational equations in the n-dimensional domain Ω and Ω^c, respectively, with coercive energy bilinear forms. For a subclass of these problems defined by a transmission problem, the coerciveness of the energy implies strong ellipticity of the corresponding boundary integral equations of the first kind (1.5.9).

Theorem 2.7 ([18]) *Let us assume that the original boundary value problem* (1.5.1), (1.5.2) *possesses an energy bilinear form $\Phi(u,v)$ with the following properties:*

For every compact $K \subset\subset \mathbb{R}^n$ there exists a constant c_K such that for all u with $u_{|\Omega} = u_{1|\Omega}$, $u_{|\Omega^c} = u_{2|\Omega^c}$ where $u_j \in C_0^\infty(K)$, $j = 1,2$, and for every v with the same properties, there hold

$$|\Phi(u,v)| \leq c_K \left(\|u\|_{H^m(\Omega)} + \|u\|_{H^m(\Omega^c)}\right)\left(\|v\|_{H^m(\Omega)} + \|v\|_{H^m(\Omega^c)}\right), \tag{2.2.13}$$

$$\begin{aligned}\text{Re}\,\Phi(u,u) &= \text{Re}\Big\{\int_\Omega \left(P^{(2m)}u\right)^T \bar{u}\,dx + \int_{\Omega^c}\left(P^{(2m)}u\right)^T \bar{u}\,dx \\ &\quad \int_\Gamma \left((R\gamma u_{|\Omega})^T\overline{S\gamma u_{|\Omega}} - (R\gamma u_{|\Omega^c})\overline{S\gamma u_{|\Omega^c}}\right) ds\Big\}.\end{aligned} \tag{2.2.14}$$

Further, there exist positive constants γ_2, ε and $c \geq 0$ such that for all u with the above properties satisfying in addition the transmission condition $R\gamma u_{|\Omega} = R\gamma u_{|\Omega^c}$, there holds the Gårding inequality

$$\text{(2.2.15)} \quad \operatorname{Re}\Phi(u,u) \geq \gamma_2 \left(\|u\|^2_{H^m(\Omega)} + \|u\|^2_{H^m(\Omega^c)}\right) - c\left(\|u\|^2_{H^{m-\epsilon}(\Omega)} + \|u\|^2_{H^{m-\epsilon}(\Omega^c)}\right).$$

Then the above assumptions imply the following Gårding inequality for the boundary integral operator A of the first kind in (1.5.9) on $\mathcal{V}^0(\Gamma)$ where

$$\mathcal{V}^\sigma(\Gamma) := \prod_{j=0}^{m-1} H^{-m+\mu_j+\sigma+\frac{1}{2}}(\Gamma);$$

i.e. there exist positive constants γ_2, ε and $c \geq 0$ such that for all $w \in \mathcal{V}^0(\Gamma)$:

$$\text{(2.2.16)} \qquad \operatorname{Re}\langle Aw, w\rangle_{L^2(\Gamma)} \geq \gamma_2\|w\|^2_{\mathcal{V}^0(\Gamma)} - c\|w\|^2_{\mathcal{V}^{-\epsilon}(\Gamma)}.$$

Consequently, in this case and in view of the Rellich imbedding theorem, the operator A can be written as

$$A = D - C$$

where D is $\mathcal{V}^0$-elliptic and C is completely continuous in $\mathcal{V}^0(\Gamma)$. Clearly, the Gårding inequality (2.2.12) or (2.2.16), respectively, implies with the Lax-Milgram theorem the invertibility of D in $H^{\frac{\alpha}{2}}(\Gamma)$ or $\mathcal{V}^0$, respectively. Hence, for strongly elliptic A including the first kind operator A in (1.5.9), there holds the *classical Fredholm alternative* due to Nikolskii's theorem [47] 5.XIII. (see also [94].)
Remark: The latter result, Theorem 2.7, for $m = 1$ can be extended to Lipschitz domains, as was shown by Costabel in [11].

2.3 Boundary element Galerkin methods

Let us introduce a family of finite-dimensional boundary element spaces $\mathcal{H}_h$ on Γ by choosing a regular $S_h^{d,r}$ family of boundary elements in the sense of Babuška and Aziz [6], $r < d$. More specifically, for $n = 2$ we use the global parametric representation (2.1.14) of Γ and introduce 1-periodic piecewise $(d-1)$-degree polynomial splines on a family of partitions $\{0 = t_0 < t_1 < \ldots < t_M = 1\}$ of the unit interval, $h := \max_{j=1,\ldots,M}\{t_j - t_{j-1}\}$. Then, with the parametric representation, we transplant the splines onto Γ, or we approximate Γ by an additional spline interpolation of the parametric representation [56]. For $n = 3$ we use a family of triangulations of the parameter domains associated with the charts of a finite atlas for Γ. On the triangulations of the parameter domains we either use pieceswise polynomial finite elements which are transplanted onto Γ by the use of the corresponding chart's application; or, in addition, we approximate Γ

by Lagrangian elements defining an approximate element surface Γ_h; and then define composite boundary elements on Γ_h by transplanting the parametric finite elements onto Γ_h following Nedelec's approach in [65].

The classical Galerkin method for solving (2.2.1) or (2.2.2) approximately uses $\mathcal{H}_h$ for the trial as well as for the test space, i.e. *find* $u_h \in \mathcal{H}_h$ *such that for all* $w_h \in \mathcal{H}_h$:

$$\langle Au_h, w_h\rangle_{L^2(\Gamma)} = \langle f, w_h\rangle_{L^2(\Gamma)}. \tag{2.3.1}$$

Let $\{\mu_j\}_{j=1}^N$ denote a Basis of $\mathcal{H}_h$. Then the desired approximate solution has the form

$$u_h(\vec{x}) = \sum_{j=1}^N \gamma_j \mu_j(\vec{x}) \tag{2.3.2}$$

where the coefficients γ_j are to be determined by solving the quadratic finite system of linear equations

$$\sum_{j=1}^N \langle A\mu_j, \mu_k\rangle_{L^2(\Gamma)}\gamma_j = \langle f, \mu_k\rangle_{L^2(\Gamma)} \quad \text{for } k = 1, \ldots, N. \tag{2.3.3}$$

These equations are equivalent to (2.3.1).

Since $A : H^{\frac{\alpha}{2}}(\Gamma) \to H^{-\frac{\alpha}{2}}(\Gamma)$, the bilinear form (2.3.1) and the influence matrix (2.3.3) will be well defined only if we require the *conformity condition*,

$$\mathcal{H}_h \subset H^{\frac{\alpha}{2}}(\Gamma) \text{ or } \mathcal{H}_h \subset \mathcal{V}^0(\Gamma), \text{ respectively.} \tag{2.3.4}$$

For the consistency of the Galerkin approximation (2.3.1) we also require the **approximation property**

$$\lim_{h\to 0} \inf_{u_h\in\mathcal{H}_h} \|w - u_h\|_{H^{\frac{\alpha}{2}}(\Gamma)} = 0; \tag{2.3.5}$$

for (1.5.9), $H^{\frac{\alpha}{2}}(\Gamma)$ is replaced by $\mathcal{V}^0(\Gamma)$.

As is well known for linear problems, the convergence $u_h \to u$ can only be established if the approximate equations (2.3.3) are stable, which can be formulated in terms of the Ladzenskaya-Babuška-Brezzi condition, in short *LBB-condition*: there exists $\gamma > 0$ such that for all $w_h \in \mathcal{H}_h$ and the whole family $\mathcal{H}_h$ with $h \to 0$,

$$\sup_{0\neq w_h\in\mathcal{H}_h} |\langle Av_h, w_h\rangle| \frac{1}{\|w_h\|_{H^{\frac{\alpha}{2}}(\Gamma)}} \geq \gamma \|v_h\|_{H^{\frac{\alpha}{2}}(\Gamma)}. \tag{2.3.6}$$

Lemma 2.8 (Céa's lemma [8], [60]) *If the LBB-condition holds then the Galerkin equations* (2.3.1) *or* (2.3.3) *are uniquely solvable and we have the quasi-optimal error estimate*

$$\|u - u_h\|_{H^{\frac{\alpha}{2}}(\Gamma)} \leq c \inf_{w_h\in\mathcal{H}_h} \|u - w_h\|_{H^{\frac{\alpha}{2}}(\Gamma)}, \tag{2.3.7}$$

where the constant c *is independent of* u *and* h.

(Throughout, c denotes a generic constant independent of the partitions and h, not the same at each occurence.)

Proof: i) Since $Av_h = 0$ implies with (2.3.6) also $v_h = 0$, and since (2.3.3) is a system of quadratic linear equations, the uniqueness implies solvability.

ii) Due to the previous arguments, the solution u_h of (2.3.3) exists for all $h > 0$ satisfying

(2.3.8) $$\langle Au_h, w_h\rangle_{L^2(\Gamma)} = \langle Au, w_h\rangle_{L^2(\Gamma)} \quad \text{for all } w_h \in \mathcal{H}_h.$$

Hence, the mapping $u \mapsto u_h := G_h u$, the ***Galerkin projection*** G_h exists for every h. Moreover, because of i), we find that G_h is a projection,

(2.3.9) $$G_h v_h = v_h \quad \text{for } v_h \in \mathcal{H}_h; \ \text{ i.e. } G_{h|\mathcal{H}_h} = I_{|\mathcal{H}_h}.$$

Moreover, for every $u \in H^{\frac{\alpha}{2}}(\Gamma)$ we have with (2.3.6)

$$\|G_h u\|_{H^{\frac{\alpha}{2}}(\Gamma)} = \|u_h\|_{H^{\frac{\alpha}{2}}(\Gamma)} \le \frac{1}{\gamma}\,|\langle Au_h, w_h^*\rangle|$$

with a specific $w_h^* \in \mathcal{H}_h$, $\|w_h^*\|_{H^{\frac{\alpha}{2}}(\Gamma)} = 1$ since on the finite-dimensional unit-sphere the supremum in (2.3.6) becomes a maximum. Inserting (2.3.8) into (2.3.9) and using continuity of the L^2-duality and of A, we find

(2.3.10) $$\begin{aligned} \|G_h u\|_{H^{\frac{\alpha}{2}}(\Gamma)} &\le \frac{1}{\gamma}\,|\langle Au_h, w_h^*\rangle| = \frac{1}{\gamma}\,|\langle Au, w_h^*\rangle| \\ &\le c\|Au\|_{H^{-\frac{\alpha}{2}}}\|w_h^*\|_{H^{\frac{\alpha}{2}}} \le c'\|u\|_{H^{\frac{\alpha}{2}}(\Gamma)} \end{aligned}$$

where the constant c' is *independent* of h and u. Hence, the LBB-condition (2.3.6) indeed implies stability.

For (2.3.7) we use the inequality

$$\begin{aligned} \|u - u_h\|_{H^{\frac{\alpha}{2}}(\Gamma)} &= \|u - w_h + G_h w_h - G_h u\|_{H^{\frac{\alpha}{2}}(\Gamma)} \\ &\le (1 + c')\|u - w_h\|_{H^{\frac{\alpha}{2}}(\Gamma)} \end{aligned}$$

for *every* $w_h \in \mathcal{H}_h$. □

Hence, the crucial property is the stability. For standard finite element methods, (2.3.6) follows directly from the coercivity of the energy form. In fact, here our situation is similar for strongly elliptic operators.

Theorem 2.9 *Let $A : H^{\frac{\alpha}{2}}(\Gamma) \to H^{-\frac{\alpha}{2}}(\Gamma)$ be a bijective, strongly elliptic $p \times p$ system of pseudo-differential operators of order α on Γ; let* (2.3.5) *and one of the following additional properties be satisfied:*

i) $\Theta =$const. *and* $C = 0$ *in* (2.2.12)*;*

ii) Θ =const.;

iii) *the boundary elements provide in addition to* (2.3.5) *the property: For every* $\varphi \in C^\infty$ *exist constants* $\delta > 0$ *and* $c_\varphi \geq 0$ *such that for all* $w_h \in \mathcal{H}_h$,

$$(2.3.11) \qquad \|\varphi w_h - P_h \varphi w_h\|_{H^{\frac{\alpha}{2}}(\Gamma)} \leq c_\varphi h^\delta \|w_h\|_{H^{\frac{\alpha}{2}}(\Gamma)},$$

where $P_h : H^{\frac{\alpha}{2}}(\Gamma) \to \mathcal{H}_h$ *denotes the* $L^2(\Gamma)$ *projection.*

Then there exists $h_0 > 0$ *such that the boundary element Galerkin method* (2.3.1) *is stable satisfying the LBB-condition* (2.3.6) *for all* $0 < h \leq h_0$. *The result also holds for* A *in* (1.5.9) *in* $\mathcal{V}^0$.

Remarks: Theorem 2.9 can be extended to more general strongly elliptic systems of equations, see e.g. [98]. The commutator property (2.3.11) can be proved for families of piecewise polynomial boundary elements on so-called K-meshes, i.e. there exists $K > 0$ such that for any two subintervals or triangles $\underset{\sim}{\triangle}$ and $\underset{\sim}{\triangle}'$ of any partition with $\overline{\underset{\sim}{\triangle}} \cap \overline{\underset{\sim}{\triangle}'} \neq \emptyset$ there holds $K^{-1} \leq |\underset{\sim}{\triangle}'|/|\underset{\sim}{\triangle}| \leq K$. Sometimes (2.3.11) is also called a super-approximation property. This property allows the application of Korn's trick, i.e. the freezing of coefficient technique, also called the localization principle, to the boundary element methods; see [5], [69].

Proof: In order to show (2.3.6) we choose any v_h and then try to find some associated w_h for which the inequality in (2.3.6) holds.

i) Here $D := \Theta A$ is $H^{\frac{\alpha}{2}}(\Gamma)$-elliptic and we can choose $w_h = \Theta^* v_h \in \mathcal{H}_h$ since here Θ is constant. Then, with $C = 0$ and (2.2.12) we find

$$\begin{aligned} |\langle A v_h, w_h\rangle| &= |\langle A v_h, \Theta^* v_h\rangle| = |\langle \Theta A v_h, v_h\rangle| \\ &\geq \operatorname{Re}\langle \Theta A v_h, v_h\rangle \geq \gamma \|v_h\|^2_{H^{\frac{\alpha}{2}}(\Gamma)}. \end{aligned}$$

With $\|v_h\|_{H^{\frac{\alpha}{2}}(\Gamma)} = \|\Theta^{*^{-1}} w_h\|_{H^{\frac{\alpha}{2}}(\Gamma)} \leq c_0 \|v_h\|_{H^{\frac{\alpha}{2}}(\Gamma)}$, $c_0 > 0$, the above inequality yields

$$|\langle A v_h, w_h\rangle| \geq \frac{\gamma}{c_0} \|v_h\|_{H^{\frac{\alpha}{2}}(\Gamma)} \|w_h\|_{H^{\frac{\alpha}{2}}(\Gamma)},$$

which implies (2.3.6).

ii) For this case, our proof follows the corresponding finite element version by A. Schatz [81]. From i) we already have for $D = \Theta A$ that the corresponding Galerkin projections

$$\underset{\sim}{G}_h w := (P_h D P_h)^{-1}_{|\mathcal{H}_h} P_h D w$$

are uniformly bounded. Moreover,

(2.3.12) $$\lim_{h\to 0} \underset{\sim h}{G} w = w \quad \text{in } H^{\frac{\alpha}{2}}(\Gamma) \text{ for every } w \in H^{\frac{\alpha}{2}}(\Gamma)$$

due to (2.3.5). Now choose

$$w_h := \Theta^* \left\{ I - \underset{\sim h}{G} D^{-1} C \right\} v_h .$$

Since for $\|v_h\|_{H^{\frac{\alpha}{2}}(\Gamma)} \le 1$ the set $D^{-1}Cv_h$ is compact in $H^{\frac{\alpha}{2}}(\Gamma)$ due to the complete continuity of C, we find from (2.3.12)

$$\|(I - \underset{\sim h}{G})D^{-1}Cv_h\|_{H^{\frac{\alpha}{2}}} \le \varepsilon(h)\|v_h\|_{H^{\frac{\alpha}{2}}}$$

with $\varepsilon(h) \to 0$ for $h \to 0$. Hence with $\Theta A = D - C$ and (2.2.12),

$$\begin{aligned}
|\langle Av_h, w_h\rangle| &= \left|\langle Av_h, \Theta^* \left\{ I - \underset{\sim h}{G} D^{-1}C\right\} v_h\rangle\right| \\
&= \left|\langle \Theta Av_h, \left\{ D^{-1}(D-C) + (I - \underset{\sim h}{G})D^{-1}C\right\} v_h\rangle\right| \\
&\ge \left|\langle \Theta Av_h, D^{-1}\Theta Av_h\rangle\right| - c\varepsilon(h)\|v_h\|^2_{H^{\frac{\alpha}{2}}(\Gamma)} \\
&\ge \operatorname{Re}\langle D(D^{-1}\Theta Av_h), (D^{-1}\Theta Av_h)\rangle - c\varepsilon(h)\|v_h\|^2_{H^{\frac{\alpha}{2}}(\Gamma)} \\
&\ge \gamma\|D^{-1}\Theta Av_h\|^2_{H^{\frac{\alpha}{2}}(\Gamma)} - c\varepsilon(h)\|v_h\|^2_{H^{\frac{\alpha}{2}}(\Gamma)} .
\end{aligned}$$

Since we assume the existence of A^{-1} and $\|w_h\|_{H^{\frac{\alpha}{2}}(\Gamma)} \le c_0\|v_h\|_{H^{\frac{\alpha}{2}}(\Gamma)}$ with $c_0 > 0$, the above estimate implies

$$|\langle Av_h, w_h\rangle| \ge (\gamma_0 - c\varepsilon(h))\, \|v_h\|_{H^{\frac{\alpha}{2}}} \|w_h\|_{H^{\frac{\alpha}{2}}}$$

with $\gamma_0 > 0$. Hence, there exists $h_0 > 0$ such that $\gamma_0 - c\varepsilon(h) \ge \frac{\gamma_0}{2}$ for all $0 < h \le h_0$ and (2.3.6) follows.

iii) Again, $D = \Theta A + C$ is $H^{\frac{\alpha}{2}}(\Gamma)$-elliptic and the corresponding Galerkin projection $\underset{\sim h}{G}$ is uniformly bounded in $H^{\frac{\alpha}{2}}(\Gamma)$. Here take

$$w_h := P_h\Theta^* \left\{ I - \underset{\sim h}{G} D^{-1} C \right\} v_h .$$

Then

$$\|w_h\|_{H^{\frac{\alpha}{2}}(\Gamma)} \le c_0\|v_h\|_{H^{\frac{\alpha}{2}}(\Gamma)} \quad \text{with } c_0 > 0$$

and

$$\begin{aligned}|\langle Av_h, w_h\rangle| &\geq \left|\langle Av_h, \left\{I - \underset{\sim h}{G} D^{-1}C\right\} v_h\rangle\right| \\ &\quad -\left|\langle Av_h, (P_h - I)\Theta^* \left\{I - \underset{\sim h}{G} D^{-1}C\right\} v_h\rangle\right| \\ &\geq \left|\langle \Theta Av_h, D^{-1}\Theta Av_h\rangle\right| \\ &\quad -\left|\langle \Theta Av_h, (I - \underset{\sim h}{G})D^{-1}Cv_h\rangle\right| - c'h^\delta \|v_h\|^2_{H^{\frac{\alpha}{2}}(\Gamma)}.\end{aligned}$$

The first two terms can be estimated in the same manner as in case ii), which yields

$$|\langle Av_h, w_h\rangle| \geq \frac{1}{c_0}\left(\gamma_0 - c\varepsilon(h) - c'h^\delta\right) \|v_h\|_{H^{\frac{\alpha}{2}}(\Gamma)} \|w_h\|_{H^{\frac{\alpha}{2}}(\Gamma)},$$

which implies (2.3.6) for all h sufficiently small.
For A in (1.5.9) in $\mathcal{V}^0$, the proof is the same as in case ii). □

Combining the previous results, we see that the Ritz-Galerkin method for injective, strongly elliptic systems of pseudo-differential or boundary integral equations under the assumptions of Theorem 2.9 is stable and converges quasi-optimally for $h \to 0$. This result can further be improved by using the finite element approximation properties of $S_h^{d,r}$-systems [6], [22], $r, d \in \mathbb{N}_0$, $r \leq d-1$.
Approximation property: *Let $\tau \leq \beta \leq d$ and $\tau < r + \frac{1}{2}$ for $n = 2$ or $\tau \leq r$ for $n = 3$. Then there exists a constant c such that for any $v \in H^\beta(\Gamma)$ there exists a sequence $\chi_h \in S_h^{d,r}$ providing*

$$\|v - \chi_h\|_{H^\tau(\Gamma)} \leq ch^{\beta-\tau}\|v\|_{H^\beta(\Gamma)}. \tag{2.3.13}$$

Theorem 2.10 ([44]) : *Let A be a bijective strongly elliptic pseudo-differential operator of order α. Let $\alpha < 2r+1$ for $n = 2$ or $\alpha \leq 2r$ for $n = 3$. Let $\alpha - d \leq \tau \leq \frac{\alpha}{2} \leq \beta \leq d$. Then we have the asymptotic error estimate of optimal order*

$$\|u_h - u\|_{H^\tau(\Gamma)} \leq ch^{\beta-\tau}\|u\|_{H^\beta(\Gamma)}. \tag{2.3.14}$$

Remarks: Costabel and Stephan present in [15] a modification of proposition (2.3.14) and its proof which allow corresponding estimates also for Petrov-Galerkin methods and for problems involving corner singularities. The extension of Theorem 2.10 to some systems of equations is straightforward [98].
Proof: i) For the special choice $\tau = \frac{\alpha}{2}$, the proposed inequality (2.3.14) follows from Céa's lemma, i.e. (2.3.7) together with (2.3.13).

ii) For $\tau \leq \frac{\alpha}{2}$ we perform the well-known Aubin-Nitsche duality arguments as in Ciarlet [9]. First observe that for a bijective pseudo-differential operator A_α on Γ also its adjoint A^* with respect to the L^2-duality is a bijective pseudo-differential operator of the same order (see [95] II.§4). We therefore know that

$$A^* : H^{\alpha-\tau}(\Gamma) \to H^{-\tau}(\Gamma)$$

is continuous and bijective; in particular, for

(2.3.15) $$A^* v = \varphi \quad \text{we have } \|v\|_{H^{\alpha-\tau}(\Gamma)} \leq c\|\varphi\|_{H^{-\tau}(\Gamma)}.$$

With the Galerkin equations we further have

$$\begin{aligned}\left|\langle u-u_h,\varphi\rangle_{L^2(\Gamma)}\right| &= \left|\langle A(u-u_h),v\rangle_{L^2(\Gamma)}\right| = \inf_{\chi_h\in\mathcal{H}_h}\left|\langle A(u-u_h),v-\chi_h\rangle_{L^2(\Gamma)}\right| \\ &\leq c\|u-u_h\|_{H^{\frac{\alpha}{2}}(\Gamma)} \inf_{\chi_h\in\mathcal{H}_h}\|v-\chi_h\|_{H^{-\frac{\alpha}{2}}(\Gamma)} \\ &\leq c' h^{\beta-\frac{\alpha}{2}}\|u\|_{H^\beta(\Gamma)}\cdot h^{\frac{\alpha}{2}-\tau}\|v\|_{H^{\alpha-\tau}(\Gamma)} \leq c'' h^{\beta-\tau}\|u\|_{H^\beta(\Gamma)}\|\varphi\|_{H^{-\tau}(\Gamma)}.\end{aligned}$$

provided $\alpha-\tau\leq d$. Here, c'' is independent of h,u,φ. Since H^τ and $H^{-\tau}$ are dual with respect to $\langle\cdot,\cdot\rangle_{L^2}$, we eventually find the proposed estimate

$$\|u-u_h\|_{H^\tau(\Gamma)} = \sup_{\|\varphi\|_{H^{-\tau}(\Gamma)}\leq 1}|\langle u-u_h,\varphi\rangle|, \leq c'' h^{\beta-\tau}\|u\|_{H^\beta(\Gamma)}.$$

□

For instance, for piecewise constant functions we have $\tau = 0$, $d = 1$, and for the boundary element Galerkin method applied to (1.2.9) where $\alpha = -1$, we find convergence of order 3,

$$|\omega-\omega_h| + \|\vec{g}_h-\vec{g}\|_{H^{-2}(\Gamma)} \leq ch^3\|\vec{g}\|_{H^1(\Gamma)},$$

see [40].

Moreover, we have the following simple consequence of Theorem 2.10.

Corollary 2.11 *Let F be a bounded linear functional on $H^{\alpha-d}(\Gamma)$ and let $u\in H^d(\Gamma)$. Then*

(2.3.16) $$|F(u)-F(u_h)| \leq ch^{2d-\alpha}\|u\|_{H^d(\Gamma)}.$$

Now let $U(x)$ denote a potential defined by the boundary charge u and $U_h(x)$ the corresponding potential defined by the charge u_h on Γ. Since for constant or C^∞-coefficients of the differential equations, the fundamental solution will be analytic or C^∞ for $x\neq y$, we find from Corollary 2.11 the pointwise estimate

(2.3.17) $$\left|\partial^l U_h(x) - \partial^l U(x)\right| \leq c\,(l,\ \mathrm{dist}\,(x,\Gamma))\, h^{2d-\alpha}\|u\|_{H^d(\Gamma)}$$

for any derivative of order l, i.e. super-convergence away from Γ. For estimates up to the boundary see e.g. the mapping properties of potentials in [21] 23.53.

In spite of the compactness of the first kind integral operators in $L_2(\Gamma)$, we could show stability and convergence of the Ritz-Galerkin method for the strongly elliptic operators in the energy space $H^{\frac{\alpha}{2}}(\Gamma)$. Nevertheless, the ill-posedness of e.g. (1.1.29) must show up in the ill-conditioning of the discrete equations (2.3.3) for $h \to 0$. For the corresponding estimates, however, we need additional regularity of the family $S_h^{d,r}$, the so-called

inverse property: *for $\tau \le \beta < r + \frac{1}{2}$ for $n = 2$ or $\tau \le \beta \le r$ for $n = 3$ there exists a constant M, such that for all $\chi_h = S_h^{d,r}$,*

$$\|\chi_h\|_{H^\beta(\Gamma)} \le M \cdot h^{\tau-\beta}\|\chi_h\|_{H^\tau(\Gamma)}. \tag{2.3.18}$$

Lemma 2.12 *Let in addition to the assumptions of Theorem 2.10 the inverse property be satisfied. Then the asymptotic estimate* (2.3.14) *holds for $\alpha - d \le \tau \le \beta \le \frac{\alpha}{2}$ and for $\frac{\alpha}{2} \le \tau \le \beta \le d$ with $\tau < r + \frac{1}{2}$ for $n = 2$ or $\tau \le r$ for $n = 3$, too. For the discrete equations* (2.3.3) *we find the estimates*

$$\begin{aligned} \|u\|_{L^2} &\le ch^{\min\{0,\alpha\}}\|P_h f\|_{L^2}, \\ \|P_h A u_h\|_{L^2} &\le ch^{-\max\{0,\alpha\}}\|u_h\|_{L^2}. \end{aligned} \tag{2.3.19}$$

Hence, the conditioning of (2.3.3) *is of order $h^{-|\alpha|}$.*

Proof: Here we will present the proof only for $|\alpha| < 2r + 1$ for $n = 2$ or $|\alpha| \le 2r$ for $n = 3$. For the remaining cases see e.g. [4].

i) $\underline{\alpha \le 0}$: Here (2.3.18) and (2.3.6) yield

$$\begin{aligned} \|u_h\|_{L^2} &\le ch^{\frac{\alpha}{2}}\|u_h\|_{H^{\frac{\alpha}{2}}} \le c'h^{\frac{\alpha}{2}}\|A^{-1}P_h f\|_{H^{\frac{\alpha}{2}}} \\ &\le c''h^{\frac{\alpha}{2}}\|P_h f\|_{H^{-\frac{\alpha}{2}}} \le c'''h^{\alpha}\|P_h f\|_{L^2} \end{aligned}$$

and coninuity with imbedding gives

$$\|P_h A u_h\|_{L^2} \le c\|u_h\|_{H^\alpha} \le c\|u_h\|_{L^2}.$$

ii) $\underline{\alpha > 0}$: (2.3.14) with (2.3.18) yields

$$\begin{aligned} \|u_h\|_{L^2} &\le \|A^{-1}P_h f\|_{L^2} + ch^{\frac{\alpha}{2}}\|A^{-1}P_h f\|_{H^{\frac{\alpha}{2}}} \\ &\le \|P_h f\|_{H^{-\alpha}} + c'h^{\frac{\alpha}{2}}\|P_h f\|_{H^{-\frac{\alpha}{2}}} \\ &\le c''\|P_h f\|_{L^2} \end{aligned}$$

and, again with (2.3.18)

$$\begin{aligned} \|P_h A u_h\|_{L^2} &\le ch^{-\frac{\alpha}{2}}\|P_h A u_h\|_{H^{-\frac{\alpha}{2}}} \le c'h^{-\frac{\alpha}{2}}\|u_h\|_{H^{\frac{\alpha}{2}}} \\ &\le c''h^{-\alpha}\|u_h\|_{L^2}. \end{aligned}$$

□

The stronger assumptions for the family of triangulations also allow pointwise error estimates.

Theorem 2.13 ([75], [76]) *For* $l \in \mathbb{N}_0$ *with* $\max\{0, \alpha - d + (3-n)/2\} \leq l \leq r$ *and under the assumptions of Theorem* 2.10 *one has the pointwise estimate,*

(2.3.20) $$\|u - u_h\|_{W^{l,\infty}(\Gamma)} \leq c|\log h|^{\frac{n-2}{2}} h^{d-l} \|u\|_{W^{d,\infty}(\Gamma)}.$$

For brevity we skip the proof.

As we have seen in this section, the strong ellipticity (2.2.10) was the decisive assumption providing stability and asymptotic convergence of Galerkin's method. The following result shows that strong ellipticity (2.2.11) is even *necessary.*

Theorem 2.14 ([82]) *Let* $n = 2$ *and let* A *be a given pseudo-differential operator of order* α *on* Γ. *Let the family of partitions provide the inverse property and assume that the boundary Ritz-Galerkin method* (2.3.1) *converges for the spline-family* $S_h^{d,d-1}$ *asymptotically with optional order* (2.3.14). *Then* A *is strongly elliptic.*

3 Collocation Methods

In engineering boundary element methods almost all the codes are based on collocation methods. Here, first choose appropriate collocation points $\vec{x}_k \in \Gamma$ associated with the family of boundary element grids, $k = 1, \ldots, N$. Then, again with (2.3.2), find the coefficients $\gamma_1, \ldots, \gamma_N$ by solving the quadratic linear system of equations,

$$Au_h(\vec{x}_k) = \sum_{j=1}^{N} A\mu_j(\vec{x}_k)\gamma_j = f(\vec{x}_k),\ k = 1, \ldots, N. \tag{3.0.1}$$

As we know from the special example in [46], the right choice of the collocation points $\vec{x}_k$ is significant for the stability and convergence of the approximation (3.0.1).

A naive choice of the $\vec{x}_k$ is an unisolvent point set for interpolation in $S_h^{d,r}$. In the following we shall present the asymptotic error analysis for two-dimensional problems where Γ is a closed curve. For $n = 3$, the collocation method was analyzed in the framework of projection methods for Fredholm integral equations of the second kind on Γ with weakly singular kernels, in particular with piece-wise constant elementes (see e.g. [33]). For the more general class of pseudo-differential operators, up to now only very few results are known; namely when Γ is a plane square, when the equations are rather specific and the boundary elements are piece-wise bilinear on square grids [87], [12].

Here, we restrict ourselves to collocation boundary element methods for two-dimensional boundary value problems, $n = 2$, where Γ is a closed curve.

3.1 Collocation with smoothest splines of piecewise odd polynomials

Here we follow the approach by Arnold and Wendland [4]. Let Γ be given by the 1-periodic global parametric representation (2.3.16). Let Δ_h denote a family of partitions of [0,1] which are 1-periodically extended to $\mathbb{R}$. Let $\{t_k\}_{k=0}^{N}$ denote the nodal points of the partition, let $\mathcal{H}_h = S_h^{r+1,r}$ be the family of $(r-1)$-times continuously differentiable splines of polynomial degree r associated with Δ_h and let r be odd, $d = r+1$, $\vec{x}_k = \vec{\Xi}(t_k)$.

Since these are the same collocation points as for stable interpolation, we call this method 'naive collocation'.

Let A be a pseudo-differential operator of order α on Γ. For continuity of Au_h at the knots $\vec{x}_k$ we assume

$$\alpha < r \ . \tag{3.1.2}$$

In the following we will show that naive collocation for strongly elliptic pseudo-differential operators is asymptotically stable and converges asymptotically with optimal order in suitable Sobolev spaces. For the special case $\alpha = 0$ and $r = 1$, Prössdorf and Schmidt showed in [71] corresponding results on uniform families of partitions Δ_h with rather different methods whereas here the partitions are arbitrary. The results presented here are optimal, the orders of convergence for naive collocation and r odd cannot be improved. They improve results in [1] and [3].

Theorem 3.1 ([4]) *The collocation equations* (3.0.1) *are equivalent to the modified Ritz-Galerkin Method: Find $u_h \in S_h^{r+1,r}$ such that for all $\chi \in S_h^{r+1,r}$*

$$((I + J_\Delta - J)\Theta(Au_h - f), \chi)_{H^\ell(\Gamma)} = 0 \tag{3.1.3}$$

where $(\cdot,\cdot)_{H^\ell(\Gamma)}$ is the scalar product (2.2.3), $\ell = \frac{d}{2} = \frac{r+1}{2}$,

$$Jw := \int_0^1 w(t)dt, \quad J_\Delta w := \frac{1}{2}\sum_{k=1}^{N} w(t_k)(t_k - t_{k-1}) \ . \tag{3.1.4}$$

Proof: Set $w = \Theta(Au_h - f)$. Then Theorem 3.1 is equivalent to the equivalence of

$$w_i := w(t_i) = 0 \text{ for } i = 1, \ldots, N \tag{3.1.5}$$

with

$$((I + J_\Delta - J)w, \chi)_{H^\ell} = (-1)^{\ell-1}\int_0^1 w'(t)\chi^{(r)}(t)dt + (J_\Delta w)(J\chi) = 0 \tag{3.1.6}$$

for all $\chi \in S_h^{r+1,r}$ where (2.2.3) was rewritten with Parseval's equality and integrated by parts $(l-1)$ times.

i) We show that (3.1.6) implies (3.1.5). First observe that the r-fold differentiation defines a bijective mapping,

$$\left(\frac{d}{dt}\right)^r : \mathcal{S}_r^0 \to \mathcal{S}_0^0 \tag{3.1.7}$$

between the $(N-1)$-dimensional spaces

$$\mathcal{S}_r^0 := \left\{\chi \in S_h^{r+1,r} | J\chi = 0\right\}$$

and

$$\mathcal{S}_0^0 := \left\{ v \in S_h^{1,0} | Jv = 0 \right\}.$$

Hence, the splines $\chi_i \in \mathcal{S}_r^0$ defined by

$$(\frac{d}{dt})^r \chi_i = v_i := \begin{cases} -h_i^{-1} & \text{for } t_{i-1} < t \le t_i, \\ h_{i+1}^{-1} & \text{for } t_i < t \le t_{i+1}, \\ 0 & \text{otherwise} \end{cases}$$

for $i = 1, \ldots, N-1$, where $h_i = t_{i+1} - t_i$, form a basis of $\mathcal{S}_r^0$. With these χ_i, (3.1.6) becomes

(3.1.8) $$(-1)^{\ell-1} \left[h_{i+1}^{-1}(w_{i+1} - w_i) - h_i^{-1}(w_i - w_{i-1}) \right] = 0$$

for $i = 1, \ldots, N-1$. Hence, exploiting (3.1.8) conscutively, we find

(3.1.9) $$h_i^{-1}(w_i - w_{i-1}) = \kappa \text{ for } i = 1, \ldots, N$$

with some fixed constant κ. Now, inserting (3.1.9) into itself with $i = N, N-1, \ldots, 1$, we find with the periodicity of w ,

(3.1.10) $$\begin{aligned} w(t_N) &= \kappa h_N + w(t_{N-1}) = \kappa(h_N + h_{N-1}) + w(t_{N-2}) \\ &= \ldots = \kappa \textstyle\sum_{j=1}^N h_j + w(t_0) \\ &= \kappa + w(t_N) \end{aligned}$$

which yields $\kappa = 0$. Therefore, (3.1.9) gives

(3.1.11) $$w(t_i) = w(0) \quad \text{for} \quad i = 1, \ldots, N.$$

Up to now, (3.1.6) was used only for $\chi \in \mathcal{S}_r^0$. But $S_h^{r+1,r} = \mathcal{S}_r^0 \oplus \{1\}$; and (3.1.6) with $\chi = 1$ yields

$$((I + J_\Delta - J)w, 1)_{H^\ell} = \frac{1}{2} \sum_{i=1}^N w_i h_i = w(0) = 0,$$

i.e. (3.1.5) because of (3.1.11). Hence,

$$\Theta(t_i)(Au_h(t_i) - f(t_i)) = 0 \quad \text{for} \quad i = 1, \ldots, N$$

and $(Det\Theta)(t_i) \neq 0$ yields (3.0.1).

ii) If (3.0.1) holds then for $w := \Theta(Au_h - f)$ we find $w_i := w(t_i) = 0$ for $i = 1, \ldots, N$ which yields

$$((I - J + J_\Delta)w, \chi_i)_{H^\ell} = 0$$

for all $\chi_i \in \mathcal{S}_r^0$. Since $J_\Delta w = 0$, we also have

$$((I - J + J_\Delta)w, 1)_{H^\ell} = 0$$

hence,

$$((I - J + J_\Delta)\Theta(Au_h - f), \chi)_{H^\ell} = 0$$

for all $\chi \in S_h^{r+1,r} = S_r^0 \bigoplus \{1\}$.

Since in our case of odd r Theorem 3.1 shows that the naive collocation method is equivalent to the modified Galerkin method (3.1.3), we can apply the techniques and results of Section 2 for proving stability and convergence. □

Theorem 3.2 ([4]) *Let A be an injective strongly elliptic pseudo-differential operator of order α on Γ and let r be odd, $r > \alpha$. Then the naive collocation method is asymptotically stable in $H^{\ell+\frac{\alpha}{2}}(\Gamma)$; i.e. there exists $h_0 > 0$ such that for all $0 < h \le h_0$ the collocation equations* (3.0.1) *are uniquely solvable and*

$$\|u_h\|_{H^{\ell+\frac{\alpha}{2}}(\Gamma)} \le c\|u\|_{H^{\ell+\frac{\alpha}{2}}(\Gamma)}. \tag{3.1.12}$$

Proof: With Theorem 3.1 we see that $u_h \in \mathcal{H}_h := S_h^{r+1,r}$ is the solution of the Galerkin method

$$(\underset{\sim}{A}(u_h - v), \chi)_{H^\ell(\Gamma)} = 0 \tag{3.1.13}$$

for all $\chi \in \mathcal{H}_h$ where

$$\underset{\sim}{A} = \Theta A \quad \text{and} \quad v = u - (\Theta A)^{-1}(J - J_\Delta)\Theta A(u_h - u). \tag{3.1.14}$$

Since $\underset{\sim}{A}$ satisfies instead of (2.2.12) the inequality

$$Re((\underset{\sim}{A} + C)w, w)_{H^\ell(\Gamma)} \ge \gamma_1 \|w\|^2_{H^{\ell+\frac{\alpha}{2}}(\Gamma)}, \tag{3.1.15}$$

a straightforward modification of the proof of Lemma 2.8 provides us with the existence of some $h_1 > 0$ and $c > 0$ such that for all $0 < h \le h_1$,

$$\|u_h - v\|_{H^{\ell+\frac{\alpha}{2}}(\Gamma)} \le c\|v\|_{H^{\ell+\frac{\alpha}{2}}(\Gamma)}. \tag{3.1.16}$$

With (3.1.14) this yields

$$\|u_h\|_{H^{\ell+\frac{\alpha}{2}}(\Gamma)} \le c\{\|u\|_{H^{\ell+\frac{\alpha}{2}}(\Gamma)} + |(J - J_\Delta)\Theta A(u_h - u)|\|\Theta A^{-1}1\|_{H^{\ell+\frac{\alpha}{2}}}\}.$$

With the error estimate for the trapezoidal rule we further estimate

$$\|u_h\|_{H^{\ell+\frac{\alpha}{2}}(\Gamma)} \le c\left\{\|u\|_{H^{\ell+\frac{\alpha}{2}}(\Gamma)} + c'h^\delta\|\Theta A(u_h - u)\|_{H^\delta(\Gamma)}\right\}$$

where $\frac{1}{2} < \delta \le min\{\ell - \frac{\alpha}{2}, 2\}$. Hence, with the continuity (2.2.9),

$$\|u_h\|_{H^{\ell+\frac{\alpha}{2}}(\Gamma)} \le c''\|u\|_{H^{\ell+\frac{\alpha}{2}}} + c'''h^\delta\|u_h\|_{H^{\ell+\frac{\alpha}{2}}}.$$

This implies (3.1.12) for $0 < h \le \min\{h_1, \frac{1}{2}c'''^{-1/\delta}\}$. □

Theorem 3.3 ([4]) *If A satisfies the assumptions of Theorem 3.2 then we have the asymptotic convergence*

$$\|u_h - u\|_{H^\tau(\Gamma)} \le ch^{\beta-\tau}\|u\|_{H^\beta(\Gamma)} \tag{3.1.17}$$

for the naive collocation, provided $0 < h \le h_0$ *and* $\alpha \le \tau \le \ell + \frac{\alpha}{2} \le \beta \le r + 1 = 2\ell$.

Remarks: Note that here no regularity assumptions are made for the family of partitions. If we compare the asymptotic convergence orders of the Galerkin method in Theorem 2.10 with that of naive collocation in Theorem 3.3 then we see that the highest possible orders are given by

$$\|u_h - u\|_{H^{\alpha-d}(\Gamma)} \le ch^{2d-\alpha}\|u\|_{H^d(\Gamma)} \quad \text{for Galerkin}$$

and

$$\|u_h - u\|_{H^\alpha(\Gamma)} \le ch^{d-\alpha}\|u\|_{H^d(\Gamma)} \quad \text{for collocation}$$

with $d = r + 1$, which is twice as high for Galerkin's method and for $\alpha = 0$ when using the same splines.

Proof: The proof is a straightforward modification of the proof of Theorem 2.10.

i) Let us denote the mapping $u \mapsto u_h$ by

$$u_h = R_h u$$

defined by solving the naive collocation equations (3.0.1) for $0 < h \le h_0$. Then R_h is a projection onto $\mathcal{H}_h$ and

$$\|R_h\|_{H^{\ell+\frac{\alpha}{2}}(\Gamma),H^{\ell+\frac{\alpha}{2}}(\Gamma)} \le c$$

uniformly due to (3.1.12), Theorem 3.2. Hence,

$$\begin{aligned}\|u - u_h\|_{H^{\ell+\frac{\alpha}{2}}(\Gamma)} &= \|u - w_h + R_h w_h - R_h u\|_{H^{\ell+\frac{\alpha}{2}}} \\ &\le (1 + c)\|u - w_h\|_{H^{\ell+\frac{\alpha}{2}}(\Gamma)}\end{aligned}$$

for *every* $w_h \in \mathcal{H}_h$ and, with (2.3.13),

$$\begin{aligned}\|u - u_h\|_{H^{\ell+\frac{\alpha}{2}}(\Gamma)} &\le (1 + c)\inf_{w_h\in\mathcal{H}_h}\|u - w_h\|_{H^{\ell+\frac{\alpha}{2}}} \\ &\le c'h^{\beta-\ell-\frac{\alpha}{2}}\|u\|_{H^\beta(\Gamma)}.\end{aligned} \tag{3.1.18}$$

ii) For $\alpha \le \tau \le \ell + \frac{\alpha}{2}$ we perform the Aubin-Nitsche duality arguments for (3.1.3). Since

$$JJ = J_\Delta J = J \ \text{ and } \ J_\Delta J_\Delta = JJ_\Delta = J_\Delta,$$

we find

$$\underset{\sim\Delta}{A}^{-1} = \underset{\sim}{A}^{-1}(I + J - J_\Delta) \text{ with } \underset{\sim\Delta}{A} := (I + J_\Delta - J)\underset{\sim}{A}.$$

Hence, the family of operators maps

$$\underset{\sim\Delta}{A}^{-1} : H^\sigma(\Gamma) \to H^{\sigma+\alpha}(\Gamma)$$

for every $\sigma \geq \frac{1}{2}$ uniformly boundedly. Correspondingly, the adjoint operators with respect to the H^ℓ-scalar product,

$$\underset{\sim\Delta}{A}^* = \underset{\sim}{A}^*(I - J + J_\Delta^*) : H^{2\ell-\sigma}(\Gamma) \to H^{2\ell-\sigma-\alpha}(\Gamma)$$

are uniformly bounded, too; as well as their inverses,

$$(\underset{\sim\Delta}{A}^*)^{-1} = (I + J - J_\Delta^*)\underset{\sim}{A}^{*^{-1}} : H^{2\ell-\sigma-\alpha}(\Gamma) \to H^{2\ell-\sigma}(\Gamma).$$

Let us consider the case $\alpha + \frac{1}{2} < \tau \leq \ell + \frac{\alpha}{2}$ first. By using duality and the Riesz representation theorem, for given $u - u_h$ we find $w \in H^{2\ell-\tau}$ such that

$$\|u - u_h\|_{H^\tau}^2 = (u - u_h, w)_{H^\ell} = \|w\|_{H^{2\ell-\tau}}^2; \tag{3.1.19}$$

and for

$$y = \underset{\sim\Delta}{A}^{*-1} w$$

we have with $\sigma = \tau - \alpha$,

$$\|y\|_{H^{2\ell+\alpha-\tau}} \leq c\|w\|_{H^{2\ell-\tau}} = c\|u - u_h\|_{H^\tau}.$$

Now (3.1.19) yields

$$\|u - u_h\|_{H^\tau}^2 = ((u - u_h), \underset{\sim\Delta}{A}^* y)_{H^\ell} = (\underset{\sim\Delta}{A}(u - u_h), (y - \chi))_{H^\ell}$$

for every $\chi \in \mathcal{H}_h$; thus

$$\begin{aligned} \|u - u_h\|_{H^\tau}^2 &\leq c\|u - u_h\|_{H^{\ell+\frac{\alpha}{2}}} \inf_{\chi\in\mathcal{H}_h} \|y - \chi\|_{H^{\ell+\frac{\alpha}{2}}} \\ &\leq c' h^{\ell+\frac{\alpha}{2}-\tau} \|u - u_h\|_{H^{\ell+\frac{\alpha}{2}}} \|y\|_{H^{\ell+\frac{\alpha}{2}-\tau}}, \end{aligned}$$

$$\|u - u_h\|_{H^\tau}^2 \leq c'' h^{\ell+\frac{\alpha}{2}-\tau} \|u - u_h\|_{H^{\ell+\frac{\alpha}{2}}} \|u - u_h\|_{H^\tau}. \tag{3.1.20}$$

Dividing by $\|u - u_h\|_{H^{\ell+\frac{\alpha}{2}}}$ and using (3.1.18) we find (3.1.17).
For the remaining case $\alpha \leq \tau \leq \alpha + \frac{1}{2}$ we set

$$y = \underset{\sim}{A}^{*^{-1}} w.$$

Then

$$\begin{aligned}\|u-u_h\|_\tau^2 &= ((u-u_h),\underset{\sim}{A}^*y)_{H^\ell}\\ &= (\underset{\sim\Delta}{A}(u-u_h),y)_{H^\ell}+(J-J_\Delta)\underset{\sim}{A}(u-u_h)\cdot Jy.\end{aligned}$$

The first term can be estimated as above. For the second term we find

$$|Jy|\le c\|y\|_{H^{2\ell+\alpha-\tau}}\le c'\|u-u_h\|_{H^\tau};$$

and with the error for the trapezoidal rule,

$$|(J-J_\Delta)\underset{\sim}{A}(u-u_h)|\le ch^{\ell-\frac{\alpha}{2}}\|u-u_h\|_{H^{\ell+\frac{\alpha}{2}}},$$

provided $\ell-\frac{\alpha}{2}\le 2$. Since here $\ell-\frac{\alpha}{2}\ge\ell+\frac{\alpha}{2}-\tau$, we have again (3.1.20) which gives (3.1.17).

In case $\alpha\le\tau\le\alpha+\frac{1}{2}$ and $\ell-\frac{\alpha}{2}>2$, we use

$$|(J-J_\Delta)\underset{\sim}{A}(u-u_h)|\le ch^2\|u-u_h\|_{H^{2+\alpha}}$$

yielding

$$\|u-u_h\|_{H^\tau}\le c_1h^{\ell+\frac{\alpha}{2}-\tau}\|u-u_h\|_{H^{\ell+\frac{\alpha}{2}}}+c_2h^2\|u-u_h\|_{H^{2+\alpha}}.$$

Here we find $\alpha+\frac{1}{2}<\alpha+2=\tau'<\ell+\frac{\alpha}{2}$. We can therefore use the previous cases for estimating the above two terms which gives (3.1.17) again. □

In this case of odd r one can even show pointwise convergence.

Theorem 3.4 ([76]) *If all assumptions of Theorem 3.2 are satisfied and, in addition, the family of partitions Δ_h is quasi-uniform then*

$$\|u-u_h\|_{C^j(\Gamma)}\le c\begin{Bmatrix}1\\|\log h|^2\end{Bmatrix}h^{d-j}\|u\|_{C^d(\Gamma)}$$
$$\textit{for } 0\le j \textit{ and } \begin{cases}\alpha+\frac{1}{2}\le j\le r & \textit{or}\\ \alpha\le j<\alpha+\frac{1}{2}.\end{cases} \tag{3.1.21}$$

For the proof see [76]. As a special example, (3.1.21) provides pointwise convergence of almost optimal order for naive collocation of strongly elliptic Cauchy singular integral equations with piece-wise linears i.e. $\alpha=j=0, d=r+1=2$.

Again, strong ellipticity is for naive collocation not only sufficient but also necessary.

Theorem 3.5 ([71], [82]) *Let A be a given pseudo-differential operator of order α on Γ. Then the naive collocation on a uniform family of partitions for smoothest splines with r odd converges asymptotically of optimal order, i.e.* (3.1.17), *if and only if A is strongly elliptic.*

3.2 Naive spline collocation for $n = 2$ on almost uniform partitions

Whereas the reduction of naive collocation (3.0.1) to Ritz-Galerkin equations via integration by parts in the previous chapter only works for odd degree splines, the collocation with even as well as with odd degree trial splines can be analyzed by using Fourier analysis and Fourier transformation of the collocation equations. The naive collocation with smoothest splines $\mathcal{H}_h = S_h^{r+1,r}$ with r even is based on the collocation points $\vec{x}_k = \vec{\Xi}(\tau_k)$ corresponding to the midpoints of the subintervals. Here we take the uniform family of partitions with knots

$$(3.2.1) \quad t_q = \frac{1}{N}(q - \frac{1}{2}), q = 1, \dots, N \text{ and } \tau_k = \frac{k}{N}, \quad k = 1, \dots, N \text{ for } r \text{ even.}$$

If r is odd then we take the collocation points at the knots,

$$(3.2.2) \qquad t_k = \tau_k = \frac{k}{N}, k = 1, \dots, N \text{ for } r \text{ odd.}$$

Of course, for smooth $\vec{\Xi}$, the points $\vec{\Xi}(t_k)$ and $\vec{x}_k$ in general will not have equal distances with respect to the arc length on Γ, but will be smoothly graded.

Here we have the following stability and convergence results.

Theorem 3.6 ([5],[79],[83]) *Let A be a strongly elliptic bijective system of pseudo-differential operators of order α. Let $\alpha < r$ if r is odd or $\alpha < r + \frac{1}{2}$ if r is even. Then the naive collocation* (3.0.1) *with smoothes splines $S_h^{r+1,r}$ is asymptotically stable and converges of optimal order, i.e. there exists $h_0 > 0$ such that for all $0 < h \le h_0$ the equations* (3.0.1) *are uniquely solvable and*

$$(3.2.3) \qquad \|u_h - u\|_{H^\tau(\Gamma)} \le ch^{\beta-\tau}\|u\|_{H^\beta(\Gamma)}$$

provided $\alpha \le \tau \le \beta \le r+1, \tau < r + \frac{1}{2}, \alpha + \frac{1}{2} < \beta$.

Remarks: Note that for r odd, here (3.2.3) also holds for $r + 1 - \alpha \le 2\beta$ exceeding the results in Theorem 3.3.

For r even and operators with even principal symbol being independent of $\vec{x}$,

$$a_0(\vec{x}; \eta) = a_0(\eta) = a_0(-\eta),$$

the estimate (3.2.3) is not sharp and can further be improved to $O(h^{r+2-\alpha})$ as was shown by Saranen in [77].

Sketch of the proof: The proof in [5] essentially consists of the following steps.

1. Prove Theorem 3.5 for equations on the unit circle with constant coefficients and operators whose symbol coincides with the principal symbol.

2. Prove Theorem 3.5 for the general equations by using Korn's trick, i.e. partition of unity on Γ, freezing coefficients and using the localization principle and perturbation arguments.

3. Extend Theorem 3.5 to smoothly graded meshes.

4. Use Aubin-Nitsche duality arguments.

For brevity, let us perform here only step 1. of the proof which was given in [79] and [5]. The collocation equations (3.0.1) can be written as

$$A_0(u_h - u)(\frac{k}{N}) = 0 \text{ for all } k \in \mathbb{Z}. \tag{3.2.4}$$

Then Fourier transformation yields

$$\sum_{j \equiv p \bmod N} \frac{1}{2}|j|^\alpha \{(1 + \mathrm{sign} j)a_0(1) + (1 - \mathrm{sign} j)a_0(-1)\} \cdot (\hat{u}_{hj} - \hat{u}_j) = 0 \tag{3.2.5}$$

for $-\frac{N}{2} < p \leq \frac{N}{2}$. The splines have N degrees of freedom, hence their Fourier coefficients satisfy the following recurrence relations for any $u_h \in S_h^{r+1,r}$ [73]:

$$\hat{u}_{h(p+kN)}(p + kN)^{r+1} = p^{r+1}(-1)^{k(r+1)}\hat{u}_{hp}. \tag{3.2.6}$$

Inserting (3.2.6) into (3.2.5) we find

$$\begin{aligned} &|p|^\alpha D(\tfrac{2p}{N})\hat{u} := \\ &|p|^\alpha \tfrac{1}{2}\{(1 + f(\tfrac{2p}{N}))(a_0(1) + a_0(-1)) \\ &+\mathrm{sign}(\tfrac{2p}{N})(1 - g(\tfrac{2p}{N}))(a_0(1) - a_0(-1))\} \cdot \hat{u}_{hp} \\ &= \sum_{j \equiv p \bmod N} (\widehat{A_0 u})_j \end{aligned} \tag{3.2.7}$$

for $0 \neq p$ and

$$\hat{u}_{h0} = \widehat{A_0 u_0} \quad \text{for } p = 0 \tag{3.2.8}$$

where

$$\begin{aligned} f(y) &= y^{r+1}|y|^{-\alpha} \sum_{k \in \mathbb{Z}\setminus\{0\}} (-1)^{k(r+1)}(\mathrm{sign} k)^{r+1}|y + 2k|^{\alpha-r-1}, \\ g(y) &= -(\mathrm{sign} y)y^{r+1}|y|^{-\alpha} \sum_{k \in \mathbb{Z}\setminus\{0\}} (-1)^{k(r+1)}(\mathrm{sign} k)^r|y + 2k|^{\alpha-r-1}, \end{aligned}$$

Both functions f and g are even, in $0 \leq y \leq 1$ they are strictly monotonically increasing, $g(1) = 1$, and

$$0 \leq 1 - g(y) \leq 1 \leq 1 + f(y) \leq c \text{ for } -1 \leq y \leq 1. \tag{3.2.9}$$

Since strong ellipticity is equivalent to (2.2.11) or to

$$\mathrm{Det}\{\frac{1}{2}(a_0(1)+a_0(-1))+\lambda\frac{1}{2}(a_0(1)-a_0(-1))\}\neq 0 \qquad \text{for all } \lambda\in[-1,+1],$$

(3.2.9) in (3.2.7) yields with $\lambda := \mathrm{sign}(\frac{2p}{N})(1-g(\frac{2p}{N}))/(1+f(\frac{2p}{N}))$ that

$$\mathrm{Det}\, D(\frac{2p}{N})\neq 0 \text{ for all } p \text{ with } -\frac{N}{2}<p\leq\frac{N}{2}.$$

Because of Lemma 2.5, there exists some constant $\gamma>0$ and a constant matrix Θ such that

$$\mathrm{Re}\,(\Theta D(y))\geq\gamma>0 \text{ for all } y\in[-1,1].$$

Hence, $D^{-1}(y)$ is *uniformly bounded*; and (3.2.7) can be solved for the Fourier coefficients of the spline solution

$$\text{(3.2.10)} \qquad \hat{u}_{hp}=D^{-1}(\frac{2p}{N})|p|^{-\alpha}\sum_{j\equiv p}(\widehat{A_0u})_j \text{ for } -\frac{N}{2}<p\leq\frac{N}{2}, p\in\mathbb{Z}\backslash\{0\}.$$

and (3.2.8). Explicit expansion of $\sum_{j\equiv p}(\widehat{A_0u})_j$ and careful estimation of the right hand side in (3.2.10) yields with (2.2.3) the desired estimate (3.2.3). For the explicit caclulations see [79] or [5]. □

Theorem 3.5 remains valid also for naive collocation with even degree splines [82], hence strong ellipticity is the decisive property for naive collocation as well as for the Ritz-Galerkin method with splines.

A different choice of the collocation points in the subintervals, however, yields asymptotically stable collocation methods converging of optimal order for a by far larger class of boundary integral equations and pseudo-differential equations on closed curves as was shown by G. Schmidt in [84].

4 Concluding Remarks

These lectures shed some light on some of the basic questions in the mathematical foundation of boundary element methods and it is quite natural that many aspects have not been presented here, although we currently witness very active mathematical research in this field.

For two-dimensional boundary value problems, the explicit Fourier analysis sketched in Section 3.2 has led to a new class of quadrature based modified collocation methods - qualocation - developed by I. Sloan [90], [91] providing higher orders of convergence than Ritz-Galerkin or collocation methods.

In order to perform any of the boundary element methods we still have to carry out numerical integration for computing the weights

$$\langle A\mu_j, \mu_k\rangle_{L^2(\Gamma)} \text{ or } (A\mu_j)(\vec{x}_k).$$

Since, in general, this cannot be done explicitly, the boundary Γ as well as these weights will be approximated and a corresponding perturbation analysis is needed. Based on Strang's lemma for bilinear forms, such an analysis was done in [65] and in [39], [57],[78]. Still, numerical integration of the above coefficients is one of the big headaches when writing boundary element codes. For two-dimensional problems, Galerkin collocation [39] provides a fully discretized numerical scheme. Combining a Taylor expansion of the kernels of the boundary integral operators with clustering of integration domains, Hackbusch and Nowak [32] combine the integration of the boundary potentials with multigrid methods for boundary integral equations [32] and develop a very fast and efficient solution procedure. (See also the recent book by Hackbusch [31].) However, most of the codes still use Gaussian elimination for solving the discrete linear systems in boundary element methods. This method becomes more efficient if it is used in combination with substructuring since the latter leads to matrices with block-banded structure and the inversion of the submatrices associated with the subdomains can be done in parallel [36]. The mathematical formulation of substructuring leads to boundary element methods for transmission problems between the substructures which also could consist of different materials [17]. If one couples finite element and boundary element methods then the advantages of each of the methods can be exploited as e.g. finite elements where one has nonlinear behavior and boundary elements for coupling with interior or exterior domains where the differential equations are linear [16], [38], [27], [68], [99]. Another coupling implants special finite elements which are modelled by a boundary element method, into a finite element formulation. Here, the special element takes care of local specialities of the geometry such as notches, cracks or corners [86]. If the geometry contains such irregularities then the solution of the boundary integral equations will develop corresponding singularities [13], [85], [96] and one needs to incorporate the singularities into the boundary element method either with mesh grading [67] or e.g., with the p-method [92].

It is quite clear that the boundary element methods will provide many new and interesting problems in the next future which hopefully will attract also some of the participants of the summer school.

Bibliography

[1] M.S. Abou El-Seoud: Kollokationsmethode für schwach singuläre Integralgleichungen erster Art. ZAMM **59** (1979) T45 - T47.

[2] M.S. Agranovič: Spectral properties of elliptic pseudodifferential operators on a closed curve. Functional Anal. Appl. **13** (1979) 279-281.

[3] M.A. Aleksidze: The Solution of Boundary Value Problems with the Method of the Expansion with Respect to Non-Orthonormal Functions. Nauka, Moscow 1978 (Russian).

[4] D.N. Arnold and W.L. Wendland: On the asymptotic convergence of collocation methods. Math. Com. **41** (1983) 383-397.

[5] D.N. Arnold and W.L. Wendland: The convergence of spline collocation for strongly elliptic equations on curves. Numer. Math. **47** (1985) 317-341.

[6] I. Babuška and A.K. Aziz: Survey lectures on the mathematical foundations of the finite element method. In: A.K. Aziz (ed.): The Mathematical Foundation of the Finite Element Method with Applications to Partial Differential Equations, pp. 3-359. Academic Press, New York 1972.

[7] C. A. Brebbia, J. C. F. Telles and L. C. Wrobel: Boundary Element Techniques. Springer-Verlag, Berlin 1984.

[8] J. Céa: Approximation variationelle des problémes aux limites. Ann. Inst. Fourier (Grenoble) **14** (1964) 345-444.

[9] P.G. Ciarlet: The Finite Element Method for Elliptic Problems. North-Holland, Amsterdam 1978.

[10] D. Colton and P. Kress: Integral Equation Methods in Scattering Theory. J. Wiley & Sons, New York 1983.

[11] M. Costabel: Boundary integral operators on Lipschitz domains: elementary results. SIAM J. Math Anal. **19** (1988) 613-626.

[12] M. Costabel, F. Penzel and R. Schneider: Error analysis of a boundary element collocation method for a screen problem in $\mathbb{R}^3$. To appear. (Preprint Technische Hochschule Darmstadt 1990.)

[13] M. Costabel and E.P. Stephan: Boundary integral equations for mixed boundary value problems in polygonal domains and Galerkin approximation. pp. 175-251 in: Mathematical Models and Methods in Mechanics 1981, W. Fiszdon and K. Wilmanski (eds.), Banach Center Publications **15** PN-Polish Scientific Publ., Warsaw, 1985.

[14] M. Costabel and E.P. Stephan: A direct boundary integral equation method for transmission problems. J.M.A.A. **106** (1985) 367-413.

[15] M. Costabel and E.P. Stephan: Duality estimates for the numerical solution of integral equations. Numer. Math. **54** (1988) 339-353.

[16] M. Costabel and E.P. Stephan: Coupling of finite and boundary element methods for an elastostatic interface problem. SIAM J. Numer. Anal. **27** (1990).

[17] M. Costabel and E. P. Stephan: Integral equations in linear elasticity. J. Integral Equations Appl. To appear.

[18] M. Costabel and W. Wendland: Strong ellipticity of boundary integral operators. Journ. Reine Angew. Math. **372** (1986) 34-63.

[19] R. Courant, D. Hilbert: Methods of Mathematical Physics II, Interscience Publ., New York 1962.

[20] Dieudonné J.: History of Functional Analysis, North Holland Amsterdam 1981.

[21] Dieudonné J.: Grundzüge der modernen Analysis VIII. Vieweg & Sohn, Braunschweig 1988.

[22] J. Elschner and G. Schmidt: On spline interpolation in periodic Sobolev spaces. P-MATH 01/83 Karl Weierstrass Inst. für Math., Akad. d. Wiss. DDR 1080, Berlin, Mohrenstr. 39, 1983.

[23] G. Eskin: Boundary Value Problems for Elliptic Pseudodifferential Equations. AMS Transl. **52**, Providence 1981.

[24] P. Filippi (ed.): Theoretical Acoustics and Numerical Techniques. CISM Courses and Lectures **277**. Springer-Verlag, Wien-New York, 1983.

[25] T. Fischer: An integral equation procedure for the exterior three-dimensional viscous flow. Integral Equations Operator Theory **5** (1982) 490-505.

[26] D. Gaier: Integralgleichungen erster Art und konforme Abbildungen. Math. Zeitschr. **147** (1976) 113-139.

[27] G.N. Gatica and G.C.Hsiao: The coupling of boundary element and finite element methods for a nonlinear exterior boundary value problem. Zeitschr. Analysis Anw. **8** (1989) 377-387.

[28] J. Giroire: L'étude de quelques problèmes aux limites éxterieurs et á résolution par équations intégrales. These d'etat. Universite de Paris VI, 1987.

[29] J. Giroire and J. C. Nedelec: Numerical solution of an exterior Neumann problem using a double layer potential. Math. Comp. **32** (1978) 973-990.

[30] N. M. Günter: Die Potentialtheorie und ihre Anwendungen auf Grundaufgaben der Mathematischen Physik. Teubner-Verlag, Leipzig 1957.

[31] W.Hackbusch: Integralgleichungen. Theorie und Numerik. Teubner-Verlag, Stuttgart 1989.

[32] W. Hackbusch and Z.P. Nowak: A multi-level discretisation and solution method for potential flow probelms in three dimsenions. Notes on Numerical Fluid Mechanics **14** (1986) 71-89.

[33] W. Hackbusch and Z.P. Nowak: On the fast matrix multiplication in the boundary element method by panel clustering. Numer. Math. **54** (1989) 463-491.

[34] F. Hartmann: The Somigliana identity on piecewise smooth surfaces. J. Elasticity **11** (1981) 403-423.

[35] F. Hartmann: Introduction to Boundary Elements. Springer-Verlag, Berlin 1989.

[36] M.F. Hodous, R.B. Katnik, D.G. Bozek and K.A. Kline: Vector processing applied to boundary element algorithms on the CDC CYBER-205. In: Vector and Parallel Computing in Scientific Applications. Pluralis, Paris 1983.

[37] L. Hörmander: Pseudo-differential operators and non-elliptic boundary value problems. Ann. of Math. **83** (1966) 129-209.

[38] G. C. Hsiao: The coupling of boundary elements and finite elements. ZAMM, to appear 1990.

[39] G. C. Hsiao, P. Kopp and W. L. Wendland: A Galerkin collocation method for some integral equations of the first kind. Computing **25** (1980) 89-130.

[40] G. C. Hsiao, P. Kopp and W. L. Wendland: Some applications of a Galerkin-collocation method for boundary integral equations of the first kind. Math. Meth. in the Appl. Sci. **6** (1984) 280-325.

[41] G. C. Hsiao and R. MacCamy: Solution of boundary value problems by integral equations of the first kind. SIAM Rev. **15** (1973) 687-705.

[42] G. C. Hsiao and R. MacCamy: Singular perturbations for the two-dimensional viscous flow problem. pp. 229-244 in: Springer Lecture Notes in Math. **942**, Springer-Verlag, Berlin 1982.

[43] G. C. Hsiao and W. L. Wendland: A finite element method for some integral equations of the first kind. J. Math. Anal. Appl. **58** (1977) 449-481.

[44] G.C. Hsiao and W. L. Wendland: The Aubin-Nitsche lemma for integral equations. J. Integral Equations **3** (1981) 299-315.

[45] G.C. Hsiao and W. L. Wendland: On a boundary integral method for some exterior problems in elasticity. Proc. Tbilisi University (Trudy Tbil. Ordena Trud. Krasn. Znam. Gosud. Univ.) UDK 539.3, Mat. Mech. Astron. **257** (1985) 31-60.

[46] V. V. Ivanov: The Theory of Approximate Methods and their Applications to the Numerical Solution of Singular Integral Equations. Noordhoff Int. Publ. Leyden 1976.

[47] L.W. Kantorowitsch and G.P. Akilow: Funktionalanalysis in normierten Räumen. Akademie-Verlag, Berlin 1964.

[48] M. Kitahara: Boundary Integral Equation Methods in Eigenvalue Problems of Elastodynamics and Thin Plates. Elsevier, Amsterdam 1985.

[49] R. Kleinman and W. Wendland: On Neumann's method for the exterior Neumann problem for the Helmholtz equation. J. Math. Anal. Appl. **57** (1977) 170-202.

[50] J. Kohn and L. Nirenberg: An algebra of pseudo-differential operators. Comm. Pure Appl. Math. **18** (1965) 269-305.

[51] J. Kral: Integral Operators in Potential Theory. Lecture Notes in Maths. **823**, Springer-Verlag, Berlin 1980.

[52] J. Kral: Boundary regularity and normal derivatives of logarithmic potentials. Proc. Royal Soc. Edinburgh **106 A** (1987) 241-258.

[53] R. Kress: Linear Integral Equations. Springer-Verlag, Berlin 1989.

[54] V. D. Kupradze: Potential Methods in the Theory of Elasticity. Israel Progr. Sci. Transl. Jerusalem 1965.

[55] V. D. Kupradze, T. G. Gegelia, M. O. Basheleishvili and T. V. Burchuladze: Three-Dimensional Problems of the Mathematical Theory of Elasticity and Thermoelasticity. North-Holland, Amsterdam 1979.

[56] M.N. Le Roux: Résolution numérique du probléme du potential dans le plan par une méthod variationelle d'éléments finis. Doctoral Thesis, University Rennes, France 1974.

[57] F. Lusikka and J. Saranen: Effect of quadratures in solving boundary integral equations by some Petrov-Galerkin methods. Preprint 1/88, Mathematics, Univ. of Oulu, Finland 1988.

[58] E. Martensen: Berechnung der Druckverteilung an Gitterprofilen in ebener Potentialströmung mit einer Fredholmschen Integralgleichung. Arch. Rational Mech. Anal. **3** (1959) 235-270.

[59] W. McLean: Periodic pseudodifferential operators and periodic function spaces. Math. Nachr., to appear. (Appl. Math. Preprint 89/10, The University of New South Wales, Australia).

[60] S.G. Michlin: Variationsmethoden der Mathematischen Physik. Akademie-Verlag, Berlin 1962.

[61] S. G. Mikhlin: Mathematical Physics, an Advanced Course. North Holland, Amsterdam 1970.

[62] S.G. Mikhlin and S. Prössdorf: Singular Integral Operators. Springer-Verlag, Berlin 1986.

[63] S. Mizohata: The Theory of Partial Differential Equations. University Press, Cambridge 1973.

[64] N. L. Muskhelishvili: Some Basic Problems of the Mathematical Theory of Elasticity. Noordhoff, Groningen 1963.

[65] J.C. Nedelec: Curved finite element methods for the solution of singular integral equations on surfaces in $\mathbb{R}^3$. Comput. Meth. Appl. Mech. Engrg. **8** (1976) 61-80.

[66] I. Netuka: Fredholm radius of a potential theoretic operator for convex sets. Časopis Pěst. Mat. **100** (1975) 374-383.

[67] T. von Petersdorff: Elasticity Problems in Polyhedra–Singularities and Approximation with Boundary Elements. Doctoral Thesis, Techn. Univ. Darmstadt 1989.

[68] C. Polizzotto: A symmetric definite BEM formulation for the elastoplastic rate problem. pp. 315-334 in: Boundary Elements IX, Vol.2 (C.A. Brebbia et al eds.) Springer-Verlag Berlin 1987.

[69] S. Prössdorf: Ein Lokalisationsprinzip in der Theorie der Splineapproximation und einige Anwendungen. Math. Nachr. **119** (1984) 239-255.

[70] S. Prössdorf and A. Rathsfeld: On strongly elliptic singular integral operators with piecewise continuous coefficients. Integral Equations Operator Theory **8** (1985) 825-841.

[71] S. Prössdorf and G. Schmidt: A finite element collocation method for singular integral equations. Math. Nachr. **100** (1981) 33-66.

[72] S. Prössdorf and G. Schmidt: A finite element collocation method for systems of singular integral equations. P-MATH 26/81. Karl Weierstrass Inst. für Math., Akad. d. Wiss. DDR 1080 Berlin, Mohrenstr. 39, 1981.

[73] W. Quade and L. Collatz: Zur Interpolationstheorie der reellen periodischen Funktionen. Sonderausgabe d. Sitzungsber. d. Preußischen Akad. d. Wiss. Phys.-math. Kl. Verlag d. Wiss. Berlin (1938) 1-49.

[74] J. Radon: Über die Randwertaufgaben beim logarithmischen Potential. Sitzungsber. Akad. Wiss. Wien, Math.-nat. Kl. Abt. IIa **128** (1919) 1123-1167.

[75] R. Rannacher and W. L. Wendland: On the order of pointwise convergence of some boundary element methods. Part I. Operators of negative and zero order. RAIRO Modél. Math. Anal. Numer. **19** (1985) 65-88.

[76] R. Rannnacher and W.L. Wendland: On the pointwise convergence of some boundary element methods. Part II: Operators of positive order. Math. Modelling Numer. Analysis **22** (1988) 343-362.

[77] J. Saranen: The convergence of even degree spline collocation solution for potential problems in smooth domains of the plane. Numer. Math. **53** (1988) 499-512.

[78] J. Saranen: On the effect of numerical quadratures in solving boundary integral equations. Notes on Numerical Fluid Mechanics **21** (1988) 196-209.

[79] J. Saranen and W.L. Wendland: On the asymptotic convergence of collocation methods with spline functions of even degree. Math. Comp. **45** (1985) 91-108.

[80] J. Saranen and W. L. Wendland: The Fourier series representation of pseudodifferential operators on closed curves. Complex Variables Theory Appl. **8** (1987) 55-64.

[81] A.H. Schatz: An observation concerning Ritz-Galerkin methods with indefinite bilinear forms. Math. Comp. **28** (1974) 959-962.

[82] G. Schmidt: The convergence of Galerkin and collocation methods with splines for pseudodifferential operators on closed curves. Zeitschr. Anal. Anwendungen **3** (1984) 371-384.

[83] G. Schmidt: On spline collocation methods for boundary integral equations in the plane. Math. Meth. Appl. Sci. **7** (1985) 74-89.

[84] G. Schmidt: On ϵ-collocation for pseudodifferential equations on a closed curve. Math. Nachr. **126** (1986) 183-196.

[85] H.Schmitz: Über das singuläre Verhalten der Lösungen von Integralgleichungen auf Fächen mit Ecken. Doctoral Thesis, Univ. Stuttgart 1989.

[86] E. Schnack: Stress analysis with a combination of HSM and BEM, pp. 273-282 in: The Mathematics of Finite Elements and Applications V (J.Whiteman ed.) Academic Press, London 1985.

[87] R. Schneider: Stability of a collocation method for strongly elliptic multidimensional singular integral equations. To appear. (Preprint 1272 Technische Hochschule Darmstadt 1989.)

[88] C. Schwab and W. L. Wendland: On the numerical integration in boundary element methods. Part I: Representations of boundary integral operators. In preparation.

[89] R. Seeley: Topics in pseudo-differential operators. pp. 169-305 In: Pseudo-Differential Operators (L. Nirenberg ed.) C.I.M.E. Lecture Notes. Editione Cremonese, Rome 1969.

[90] I.H. Sloan: A quadrature-based approach to improving the collocation method. Numer. Math. **54** (1988) 41-56.

[91] I.H. Sloan and W. L. Wendland: A quadrature based approach to improving the collocation method for splines of even degree. Zeitschr. Anal. Anwendungen **8** (1989) 361-376.

[92] E.P. Stephan and M. Suri: On the convergence of the p-version of the boundary element Galerkin method. Math. Comp. **52** (1989) 31-48.

[93] E.P. Stephan and W.L. Wendland: Remarks to Galerkin and least squares methods with finite elements for general elliptic problems. pp. 461-471 in: Partial Differential Equations. Lecture Notes in Mathematics **564**, Springer-Verlag, Berlin 1976 and in Manuscripta Geodaetica **1** (1976) 93-123.

[94] F. Stummel: Rand- und Eigenwertaufgaben in Sobolewschen Räumen. Lecture Notes Math. **102**, Springer-Verlag, Berlin 1969.

[95] M. E. Taylor: Pseudodifferential Operators. Princeton Univ. Press, Princeton 1981.

[96] K. Volk: Zur Berechnung von Singulärfunktionen dreidimensionaler elastischer Felder. Doctoral Thesis, Univ. Stuttgart 1989.

[97] W. L. Wendland: Assymptotic accuracy and convergence for point collocation methods. In: C. A. Brebbia (ed.): Topics in Boundary Element Research Vol. 3 Chap.9, pp. 230-257. Springer-Verlag, Berlin 1985.

[98] W. L. Wendland: Strongly elliptic boundary integral equations. In: A. Iserles, M. Powell (eds.): The State of the Art in Numerical Analysis, pp. 511-561, Clarendon Press, Oxford 1987.

[99] W.L. Wendland: On asymptotic error estimates for the combined BEM and FEM. pp. 273-333 in: Finite Element and Boundary Element Techniques from Mathematical and Engineering Point of View. (E.Stein, W. Wendland eds.) CISM Lecture Notes **301**, Springer-Verlag, Wien-New York 1988.

GPSR Compliance

The European Union's (EU) General Product Safety Regulation (GPSR) is a set of rules that requires consumer products to be safe and our obligations to ensure this.

If you have any concerns about our products, you can contact us on ProductSafety@springernature.com

In case Publisher is established outside the EU, the EU authorized representative is:

Springer Nature Customer Service Center GmbH
Europaplatz 3
69115 Heidelberg, Germany

Batch number: 10577876

Printed by Printforce, the Netherlands